Jodi Bassett is an Australian writer, fine artist, designer, patient advocate, and the founder of the international M.E. charity, The Hummingbirds' Foundation for Myalgic Encephalomyelitis (HFME).

Jodi contracted M.E. in 1995 when she was just 19. Due largely to misdiagnosis and inappropriate medical advice in the early stages of the disease, and because she was not told of the importance of avoiding overexertion in M.E., she is currently severely affected, housebound and largely bedbound.

For the same or similar reasons, the majority of HFME contributors are likewise disabled.

Very little advocacy exists for M.E. patients, and HFME contributors have determined that despite their high disability levels, they must do what they can for M.E. rights.

HFME contributors also aim to advocate for those non-M.E. patients who have been given the always meaningless 'Chronic Fatigue Syndrome' diagnosis, and subsequently denied correct diagnosis and treatment.

BOOKS BY JODI BASSETT

Caring for the M.E. Patient

COMING SOON FROM THE SAME AUTHOR

'CFS' – An Illogical, Harmful and Potentially Deadly Misdiagnosis (working title)

The HFME Guide to Basic M.E. Treatment (working title)

Overexertion and the M.E. Patient (working title)

What M.E. Feels Like (working title)

M.E.: A Guide for Doctors (working title)

What is M.E.? (booklet)

CARING FOR THE M.E. PATIENT

A HFME GUIDE

Essential information for anyone who knows, loves
or provides care for someone with M.E.,
sourced from the world's leading M.E. experts.

FIRST EDITION. PUBLISHED 2011. JODI BASSETT.

This book is dedicated to you,
the reader

A FOREWORD BY DR BYRON HYDE

Unfortunately, I have not met Jodi Bassett of the Hummingbirds Foundation for M.E. although in September 2010 I came close to visiting her home when I was lecturing in the incredibly beautiful city of Perth in Western Australia. Nor have I met her charming mother or her father who have both been indispensible in assisting Jodi and her work.

Jodi is a very courageous young woman who not only runs one of the few balanced on-line M.E. world newsletters and websites but recently has completed a serious book on the subject of M.E. This is a book that deserves being read, not only by patients and physicians with an interest in M.E. but the bureaucrats in the USA Centers for Disease Control who have done so much damage to the understanding of the M.E. illness spectrum.

It is not just very difficult, it is almost impossible to hold a balanced and intelligent view on a subject such as M.E. without spending one's life actually examining and investigating the patient. Putting the name of a disease on a patient is never enough; what is essential is the proper examination of a disabled patient to understand how as a physician the patient may be assisted.

There is also so much false information that is picked up and disseminated by the M.E. and CFS press it is near impossible to hold one's head above the water and sift through this morass of misinformation. Much of the misinformation on M.E. and CFS is published by individuals and companies who own test laboratories or who hold patents on viruses or viral evaluation. Any attempt to seek the truth is always a major difficulty.

Somehow, Jodi Bassett and Hummingbird have managed to plow through this field of weeds. I recommend her book to all and wish it every best success.

Byron M. Hyde M.D.

Byron Marshall Hyde MD
Sicily, August 2011

How to use this book

For friends, family and partners:
- To get the basic facts about M.E. quickly, read one or all of the three brief papers in Chapter 1.
- To learn how you can best support the M.E. patient in your life, read "So you know someone with M.E." on page 25.
- The first half of Chapter 3 is recommended where the person you know has severe M.E
- To learn as much as possible about M.E., read the entire book.

For carers of M.E. patients:
- To get the basic facts about M.E. quickly, read one or all of the three brief papers in Chapter 1.
- To learn how to provide care for an M.E. patient, read "Hospital or carer notes for M.E." on page 41.
- Chapter 3 is particularly recommended for carers.
- To learn as much as possible about M.E., read the entire book.

For doctors:
- Doctors interested in the correct diagnosis, management and treatment of M.E. are advised to read "What is M.E.?" on page 113 and "Hospital or carer notes for M.E." on page 41.
- Also recommended reading are all of the papers in Chapters 1, 3 and 4.
- Other relevant papers may be found at the HFME website (www.hfme.org). The most important papers on the HFME website for doctors are: *The misdiagnosis of 'CFS,' M.E. is not fatigue or 'CFS,' Testing for M.E. (and Dr Byron Hyde's 2007 Nightingale definition of M.E.), The myths of M.E., The importance of avoiding overexertion in M.E.* and *Treating M.E. – The basics*

For M.E. patients:
- This book makes a very suitable primer for the M.E. patient. Patients may be best served by first reading this book themselves (or as much of it as is possible, beginning with Chapter 1) and then passing it on to friends and family members.
- Additional papers for M.E. patients can be found at the HFME website (www.hfme.org).
- Patients may also wish to fill out some or all of the four forms provided in this book, in order to give those friends and family members (or medical staff) additional information about how M.E. affects them personally. This can be done by writing directly inside the book (where appropriate), on a photocopy taken from the book or a printout of the free downloads available from the website.

  Patients may then like to call attention to the filled-in forms to the intended reader with a note attached to the front of the book or by using sticky flags to mark the appropriate individual pages.

  The four forms in this book are on the following pages:

  - Page 36: The misdiagnosis letter for M.E. patients: Form
  - Page 47: Hospital or carer notes for M.E. patients: M.E. patient's care form
  - Page 96: The HFME ability scale for M.E. patients: Summary
  - Page 98: The HFME ability and severity checklist

Contents

Notes on the formatting, design & use of this book

Before reading this book, please note:

1. As some visitors to the HFME website (www.hfme.org) or readers of this book may only ever read one HFME paper, each paper has been designed to be a stand-alone resource which focuses on one aspect of Myalgic Encephalomyelitis (M.E.) but which also includes a brief rundown of the basic facts of M.E. Thus there is significant repetition of the basic facts of M.E. (and related topics) from paper to paper in this book.

If you have read the basic facts once and have no need to be reminded of this information again, please just skim over the repetitive sections when you encounter them in future papers.

2. The papers in this book were originally created to be published online, and distributed for free, on the HFME website. There are many small differences in how information is presented online and in print form. In an ideal world each HFME paper would have been completely reformatted and reorganised, before being included in this book. Unfortunately, due to the serious illness and disability suffered by the author/s, total reformatting and reorganising of each paper was not possible.

Thus this book includes some minor formatting inconsistencies. Where further information is recommended, the links given are in an online format (i.e. HTML links appear here as underlined text). There are also almost certainly some minor grammatical errors.

However, we ask readers to ignore these superficial imperfections and to focus on the far more important fact that the information given in this book on M.E. (and related topics) is rock solid. It has been compiled using information from the world's leading M.E. experts – and a large number of M.E. patient accounts spanning many decades – and is of the highest quality. This is information that is currently unknown by most of the public, the media, doctors and even patients themselves, and that desperately needs to become known – this is why those involved with HFME have produced this book, despite their serious illness and disability caused by M.E.

3. To be able to follow any of the 'links' to further information given in this book (represented by underlined text), just go to the HFME website, view the online version of the relevant paper, and click on the relevant link.

4. If you would like printouts of any of the papers in this book for yourself or to hand out to doctors or others, you can download free printable copies of each paper in this book from the HFME website. See the 'Document downloads' page on the website for more information.

Permission is given for all HFME papers in this book to be freely redistributed by email or in print for any not-for-profit purpose provided that the entire text (including the copyright notice, the author's attribution and the HFME logo) is reproduced in full and without alteration. Knowledge is power. Please redistribute these texts, and the HFME books, widely.

5. To learn more about HFME and the aims of HFME, please see Chapter 5 of this book.

6. To read the full reference list for each HFME paper, please see Chapter 5 of this book.

CHAPTER ONE
Introduction

The basic facts of Myalgic Encephalomyelitis (M.E.) cannot be explained in a mere two or three sentences, but these facts are not time consuming to take in or impossibly difficult to understand. This is particularly true when M.E. is separated out from the vague, illogical and designed-to-be-confusing mess that is 'CFS' (Chronic Fatigue Syndrome).

This book will start with the basics, which are:

- To describe briefly what M.E. is,
- To explain some of the terrible problems that people with M.E. are facing, and why they so badly need your support and understanding,
- To provide a brief explanation of *why* these problems are happening and why they continue to occur.

These fundamental facts and issues will help you better understand M.E. and those who suffer from it.

This introductory chapter includes the following three papers:

1. The one page summary of the facts of M.E.

2. M.E.: The shocking disease.

3. What is M.E.? Summary

A one-page summary of the facts of M.E.

- Myalgic Encephalomyelitis (M.E.) is a disabling neurological disease that is very similar to Multiple Sclerosis (M.S.) and Poliomyelitis. Earlier names for M.E. were 'atypical Multiple Sclerosis' and 'atypical Polio.'

- M.E. is a neurological disease characterised by scientifically measurable post-encephalitic damage to the brain stem. This damage is an essential part of M.E., hence the name M.E. The term M.E. was coined in 1956 and means: my = muscle, algic = pain, encephalo = brain, mye = spinal cord, tis = inflammation. This neurological damage has been confirmed in autopsies of M.E. patients.

- Myalgic Encephalomyelitis has been recognised by the World Health Organisation's International Classification of Diseases since 1969 as a distinct organic neurological disease. M.E. is classified in the current WHO International Classification of Diseases with the neurological code G.93.3.

- M.E. is primarily neurological, but also involves cognitive, cardiac, cardiovascular, immunological, endocrinological, metabolic, respiratory, hormonal, gastrointestinal and musculo-skeletal dysfunctions and damage. M.E. affects all vital bodily systems and causes an inability to maintain bodily homeostasis. More than 64 individual symptoms of M.E. have been scientifically documented.

- M.E. is an acute (sudden) onset, infectious neurological disease caused by a virus (a virus with a 4-7 day incubation period). M.E. occurs in epidemics as well as sporadically and over 60 M.E. outbreaks have been recorded worldwide since 1934. There is ample evidence that M.E. is caused by the same type of virus that causes Polio; an enterovirus.

- M.E. can be more disabling than M.S. or Polio, and many other serious diseases. M.E. is one of the most disabling diseases that exists. More than 30% of M.E. patients are housebound, wheelchair-reliant and/or bedbound and are severely limited with even basic movement and communication.

- *Why are M.E. patients so severely and uniquely disabled?* For a person to stay alive, the heart must pump a certain base-level amount of blood. Every time a person is active, this increases the amount of blood the heart needs to pump. Every movement made or second spent upright, every word spoken, every thought thought, every word read or noise heard requires that more blood must be pumped by the heart.

  However, the hearts of M.E. patients only barely pump enough blood for them to stay alive. Their circulating blood volume is reduced by up to 50%. Thus M.E. patients are severely limited in physical, cognitive and orthostatic (being upright) exertion and sensory input.

  This problem of reduced circulating blood volume, leading to cardiac insufficiency, is why every brief period spent walking or sitting, every conversation and every exposure to light or noise can affect M.E. patients so profoundly. Seemingly minor 'activities' can cause significantly increased symptom severity and/or disability (often with a 48-72 hour delay in onset), prolonged relapse lasting months, years or longer, permanent bodily damage (e.g. heart damage or organ failure), disease progression or death.

  If activity levels exceed cardiac output by even 1%, death occurs. Thus the activity levels of M.E. patients must remain strictly within the limits of their reduced cardiac output just in order for them to stay alive. *M.E. patients who are able to rest appropriately and avoid severe or prolonged overexertion have repeatedly been shown to have the most positive long-term prognosis.*

- M.E. is a testable and scientifically measurable disease with several unique features that is not difficult to diagnose (within just a few weeks of onset) using a series of objective tests (e.g. MRI and SPECT brain scans). Abnormalities are also visible on physical exam in M.E.

- M.E. is a long-term/lifelong neurological disease that affects more than one million adults and children worldwide. In some cases M.E. is fatal. (Causes of death in M.E. include heart failure.)

M.E. - The shocking disease

 In thinking about M.E. and all of the terrible things that are happening so unfairly to so many wonderful innocent people year after year, and how extremely severe a disease it can be physically, I keep coming back to one word: shocking. These are the basic M.E. facts:

- M.E. is similar in significant ways to illnesses such as Multiple Sclerosis (M.S.), Lupus and Polio.

- M.E. occurs in epidemic and sporadic forms, over 60 M.E. outbreaks have been recorded worldwide since 1934.

- What defines M.E. is a specific type of acquired damage to the brain (the central nervous system) caused by a virus (an enterovirus). It is an *acute (sudden) onset* neurological disease initiated by a virus infection with multi system involvement which is characterised by post encephalitic damage to the brain stem.

- The term M.E. was coined in 1956 and means: My = muscle, algic = pain, encephalo = brain, mye = spinal cord, itis = inflammation. This neurological damage has been confirmed in autopsies of M.E. patients.

- M.E. is primarily neurological, but also involves cognitive, cardiac, cardiovascular, immunological, metabolic, respiratory, hormonal, gastrointestinal and musculo-skeletal dysfunctions and damage. M.E. causes an inability to maintain bodily homeostasis. More than 64 individual symptoms of M.E. have been scientifically documented.

- M.E. can be more disabling than M.S. or Polio, and many other serious diseases. M.E. is one of the most disabling diseases there is. More than 30% of M.E. patients are housebound, wheelchair-reliant and/or bedbound and are severely limited with even basic movement and communication. In some cases M.E. is fatal.

- The hearts of M.E. patients barely pump enough blood for them to stay alive. Their circulating blood volume is reduced by up to 50%. Thus M.E. patients are severely limited in physical, cognitive and orthostatic (being upright) exertion and sensory input. This problem of reduced circulating blood volume, leading to cardiac insufficiency, is why every brief period spent walking or sitting, every conversation and every exposure to light or noise can affect M.E. patients so profoundly.

    Seemingly minor 'activities' can cause significantly increased symptom severity and/or disability (often with a 48-72 hour delay in onset), prolonged relapse lasting months, years or longer, permanent bodily damage (e.g. heart damage or organ failure), disease progression or death.

    If activity levels exceed cardiac output by even 1%, death occurs. Thus the activity levels of M.E. patients must remain strictly within the limits of their reduced cardiac output just in order for them to stay alive.

    M.E. patients who are able to rest appropriately and avoid severe or prolonged overexertion have repeatedly been shown to have the most positive long-term prognosis.

- M.E. is a testable and scientifically measurable disease with several unique features that is not difficult to diagnose, even within just a few weeks of onset, using a series of objective tests.

- M.E. is a debilitating neurological disease which has been recognised by the World Health Organisation (WHO) since 1969 as a distinct organic neurological disorder. M.E. is classified in the current WHO International Classification of Diseases with the neurological code G.93.3.

- Many patients with M.E. do not have access to even basic appropriate medical care. Medical abuse and neglect is also extremely common and often results in the disease becoming severe (and in some cases, death is caused).

- Governments around the world are currently spending $0 a year on M.E. research.

These facts however, fall far short of getting across what a hell on earth M.E. really is. Above all else, I think M.E. is a shocking disease. These are a few of the biggest shocks I've faced, and that others with M.E. also experience:

1. The shock of extremely severe sudden illness and disability

The first big shock is how quickly and completely your entire life can change forever. Having your body suddenly act very differently and not be able to do all the things you have done many thousands or millions of times before, is surreal. This is especially so when this occurs suddenly from one day to the next, as it does with M.E. The sense of unreality can be so strong that you almost wonder why everyone else is still going on as if nothing had changed and everything was normal.

For me, in March 1995 at the age of 19, I went from being very healthy and happy one day to having problems standing upright for more than a few minutes at a time, the next. I also suddenly had severe problems sleeping, thinking and remembering, speaking and understanding speech, eating many foods that I previously tolerated perfectly well, coping with even low levels of noise and light and vibration, coping with warm weather, sitting, with my heart and blood pressure, with any type of physical or cognitive activity causing severe relapse unless within very strict limits as well as memory loss, facial agnosia, learning difficulties, severe pain, alcohol intolerance, blackouts and seizures, intense unusual headaches, burning eyes and ear pain, rashes (and other skin problems), severe nausea and vertigo, total loss of balance when I closed my eyes or the room was dark, muscle weakness and paralysis, and so on.

I suddenly had over 60 individual symptoms, and could only do 40% or less of my pre-illness activities.

It's a bit like one day waking up and suddenly everyone around you is speaking another language and looking at you strangely for not being able to understand what is being said.

At first, not only is it very hard to just accept, but also to really believe it is happening, and that it won't just go away as suddenly as it came. It's all just such a big shock.

2. The medical system shock

As if that weren't enough all on its own, the next big shock involves lifelong beliefs about our medical system. You soon find out that the disease you have is one of those that is treated differently from many others, that not every disease is viewed equally, and that bizarrely this has *nothing at all* to do with the type of disease or the severity of the disease or its symptoms, or testable abnormalities, or the possibility of death, but other non-scientific and non-medical factors. It has to do with political and financial factors, and marketing.

You find out that some diseases get you 'red carpet' treatment or and guarantee that you are treated very well, others are treated adequately, and unfortunately several leave you with no real care at all. Even worse, some diagnoses subject you to serious mistreatment from the professionals *meant to be there to help you.*

Most people trust absolutely that if they get severely ill, they can go to an emergency room and be given the appropriate medical care. I used to trust in that too. But I was to soon find out the hard way that that didn't apply to me anymore. If I went to the emergency room, there was an enormous chance I'd not only get no help at all, but be ridiculed into the bargain or told 'to stop exaggerating' or refused the appropriate tests (and have older test results ignored). I may then be told, illogically, and despite all the evidence to the contrary that 'there is nothing wrong with you, go home and let us care for someone who is really ill.' I'd be far more likely to come out of the emergency room far sicker than when I'd gone in (in crisis), as well as being verbally abused and insulted as well.

Dealing with GPs and specialists is much the same most of the time, for those with M.E. Probably the most common treatment recommended to patients with M.E. is graded exercise therapy (GET) (both formal and informal programs). This is a 'treatment' that can and very often does leave M.E. patients, including children, far sicker afterwards for months, years or longer (wheelchair-reliant, bedbound, needing intensive care etc.). It can also cause death. While it may help some of those with other illnesses very different to M.E., it has a zero percent chance of providing any benefit to M.E. patients. If even a tiny percentage, say 2%, of almost any other patient group were made as ill and disabled by any treatment (as M.E. patients are by GET) it would be a huge scandal. It would make all the papers and there would be all sorts of legal actions and enquiries, and outpourings of public outrage. Yet the incidence of M.E. patients being recommended, or forced or coerced, into this torture is growing every year. Nobody much cares or even knows. It's more than shocking or just very cruel, it's obscene.

Most people have no idea that all this medical abuse occurs regularly, to people *just as ill or even far more ill* as those with M.S. or Lupus. When you do try to tell them, most often they refuse to believe it could be true, so strong is their belief in the fairness and logic of our health system and how much thought, objectivity and careful investigation supposedly goes into giving a final diagnosis and recommending treatment. It's a shocking loss, this loss of belief in a health safety net and a medical system based on logic, science and due care. It's such a comforting belief, it's hardly surprising people don't want to give it up, even if it is false.

Thanks to inappropriate medical care, I, as with many other M.E. patients, soon struggled to do even 5% of the activities I had pre-illness. I was made housebound and 99% bedbound, and have remained so for the last 10 years. My heart-rate skyrockets and my blood pressure drops dramatically after just a few minutes of standing or other overexerting activity. It feels like a heart attack in every organ, and as if my heart is about to explode, or just stop. (The highest heart-rate measurement I've had is 170 bpm and the lowest blood pressure measurement is 79/59 – both were taken at times when I was only moderately ill, relatively speaking, nowhere near my most severe state. Scary.)

I have spent most of the last decade, alone and in pain in a dark quiet room, coping with many different and hideous symptoms. I accept that some people get ill, and that I am at risk of this as much as anyone. What is hard to take is that, like so many M.E. patients, my reaching such a severe disability level and losing so much of my life was completely unnecessary and would very likely not have happened had I had even the most basic appropriate support in the beginning.

3. The welfare system shock

Despite being extremely ill and disabled, M.E. patients are often shocked to find that getting the basic welfare payments is very difficult or impossible. Bizarrely enough, the system is set up in such a way that you can actually be *too ill* to qualify, as so many hoops are required to be jumped through to lodge a successful claim, without which the claim is denied. Ironically, the government agencies seem to have little interest in this conundrum, nor in how much sicker jumping through all their hoops makes you long-term. The ignorance of doctors and their inability to give you an unbiased examination is also a huge problem.

Again, what is far more important to them is the name and reputation of your disease, not how ill and disabled you are. It is not uncommon to find instances of M.E. patients living for years with no disability payments, having to live on the mercy of family, or becoming homeless.

4. The media shock

The public largely trusts the information given about different diseases in the media. I did too, and I still do largely, provided the article is about M.S. or cancer. But like many M.E. patients, I was shocked to find out that when it came to diseases like mine, there was no onus at all on the reporter to be accurate. While a furore would ensure if articles made up *entirely* of false information were printed about M.S. or cancer, almost every article that I read about M.E. was of an unbelievably low standard, yet nobody seemed to care at all.

Similarly, the outrage when certain groups are made fun of in what is deemed an offensive manner, simply does not occur when it's M.E. that is being ridiculed. For some reason M.E. patients (in the UK particularly) are fair game. This is because despite the fact that our governments have created laws designed to stop discrimination on the basis of gender, race and disability and so on, discrimination against M.E. patients is not only allowed, but is actively supported and promoted by government. (For information on *why* this occurs, see What is M.E.?)

5. The human rights groups shock

While the big human rights groups seem very eager to help many other groups and even individuals facing small or large problems, they seem completely unwilling to even look at the severe abuse of human rights facing perhaps a million M.E. patients worldwide, let alone do anything at all to actually help. This when even the smallest action on their part, the smallest indication of support for the M.E. cause, would be such a huge step forward for M.E. patient rights. Such ignorance and injustice is shocking.

6. The friends and family shock

What makes coping with all these things unimaginably worse is having to do so with little if any support from friends and family – and even while facing abuse or ridicule from friends and family. Some patients are even disowned by their whole family, or all but a few members.

Loved ones often believe the same financially-motivated media and government-sanctioned nonsense about your disease as the doctors do. They often accept the 'miracle cure' stories in the media featuring people with a hundred different mild (and sometimes psychological) or transient diseases jumping up and down about how they have been 'cured' by the mumbo jumbo money-making scam of the week – despite the fact that none of these stories features actual M.E. patients, or even patients with diseases similar to M.E.

It's such a huge shock that those you love could see you so ill and refuse to support you and that they have more trust in doctors than in your integrity. They can't believe that if you were seriously ill, a doctor could miss it, even though that is exactly what has happened. They can't believe that the media would be allowed to print completely fictional information about your disease often based on mixed and *entirely unrelated* patient groups, even though they *are* doing just that. Not having medical or media (or government) support makes getting support from loved ones almost impossible.

Having loved ones not stand by you hurts a lot, in many ways. It takes yet another huge swipe at what self-esteem you have left after being treated like dirt by your trusted doctors and welfare departments, leaving study incomplete and/or losing your job and your ability to support yourself and/or being denied the services of a carer when you urgently need one. After so many attacks on your integrity and worth you can't help but be worn down by it all, particularly when you're so ill and even more so if you are not yet of adult age when you become ill. You inevitably feel, not depressed, but as if you must personally be unworthy somehow of any type of care or compassion. Such messages inevitably sink in to some extent after constant repetition, no matter how educated, strong or mentally fit you are.

7. The M.E. charities and support groups shock

Realising that very nearly all of the charities and support groups that claim to be there to help you actually do not represent or support you at all and are actually hostile to your interests is yet another huge shock.

You go to a group that you trust finally to give you the unadulterated facts and to be working towards improving things and all you get is more abuse and misinformation. Just as bad, you also don't get all the important information about M.E. that could make an enormous positive difference to your life and to your health. If you try to improve matters and provide these groups and individuals with accurate information you are either ignored or banned, perceived as negative.

M.E. patients are in a terrible position. Almost all 'our' charities have sold themselves off to the highest bidder, and are now working to promote the same harmful misinformation they were created to fight against.

(The concepts of 'CFS' and 'ME/CFS' can be immensely profitable to some groups, as is explained in several other HFME papers.) These groups claim to be representing a large and diverse patient group but in reality they do not work for the benefit of any group, except themselves. They often take advantage of patients' lack of ability (or unwillingness) to engage with politics, to read and assimilate significant amounts of slightly complex text and of their goodwill and trust, in the cruellest way. Many patients put all their faith and efforts into this false advocacy, led by vested interest groups. Many (perhaps even most) fellow patients are, unwittingly, working directly against their own interests and aiding their abusers. Many seem determined to support the same old illogical nonsense that is the reason that no progress at all has been made in over 20 years. Perhaps some patients are too ill to even investigate other sources of information than the charity, or they have taken the charity's word for it that the (entirely bogus) information they provide is all that exists.

These sell-out groups and individuals are at fault here to a large extent, but at the same time they couldn't keep doing the evil things they do if they didn't have so much undeserved (and extremely unwise) patient support. It's so incredibly shocking, and *frustrating.*

Those few groups and individuals that are involved in genuine advocacy are often able to do very little due to the physical constraints of M.E., the poverty associated with M.E., and the lack of public and other support. M.E. patients are just too ill to fight effectively for themselves like AIDS patients did. They can't rally or march and many can only barely read or write now. AIDS patients also often have an early asymptomatic period of illness, which enables significant contribution to activism – but for M.E. patients the severe symptoms begin on day one.

8. The M.E. advocacy nightmare shock

Perhaps most shockingly of all, when you try to do some advocacy yourself and tell people about the double standards, discrimination and unfair treatment, and show them mountains of solid facts, you are met with disbelief. People cannot or will not believe that doctors could be so cruel, unscientific, ignorant and illogical; or that our governments and media could be so unethical and dishonest by selling their integrity for political and financial gain; or that so-called 'charities' could be just as corrupt.

Many people refuse to even do a tiny bit of quality reading on the topic of M.E., wrongly believing that they already have all the facts and know all there is to know, believing that anything that they don't know just can't be true. If you try to give people correct information you are accused of exaggerating or being fanciful. People snicker or roll their eyes when you talk about cover-ups, and give your information as much credence as stories of alien abductions or the 'false' moon landing. Anything not already mainstream is met with scepticism, as is the idea that all of these groups could *collaborate* to create a mutually profitable, and very hard to undermine, lie.

Despite ample evidence of similar scandals and cover-ups in the past, people seem unwilling to give up their belief in a fair and just government, media and medical system. They refuse to give up their comforting delusions...until and unless something similar happens to them, at least, and they have no choice but to face reality. But then, of course, they too are disbelieved when they try to spread the word, and so on and on it goes.

Most families and friends of patients are completely unwilling to help with advocacy, very often due to ignorance about the medical and political facts of M.E. Others are too busy with the duties of a carer for advocacy. Patients with other diseases almost always do not understand that the most commonly given information on M.E. is entirely false. By believing M.E. is something it is not and reinforcing many of the worst myths about the disease, most of these well-meaning groups and individuals work directly against the interests of M.E. patients, sadly.

M.E. itself also seems to work against you, in an unexpected way. People say it's too severe and there are too many symptoms. The entirely unique way we respond to even trivial exertion and are so disabled by it, instead of inspiring sympathy, seems to actually inspire disbelief. People seem to (bizarrely) believe that there must be some limit on how bad a disease could be, and that such severe illness couldn't be possible long-term. That you couldn't possibly be too ill to sit or stand up, use the phone, speak or be spoken to, listen to music, write a letter, spend time in hospital or take a short trip out of the house; that you couldn't possibly be so ill that you can only dream of one day being *well enough* to use an electric wheelchair sometimes, if you're really lucky – and so on. As if all humans were 'guaranteed' somehow to always be able to at least do such simple tasks, and to only ever suffer a 'reasonable' level or time period of disability. But the body does not acknowledge such limits. If only.

In 20 years not only has no progress been made in the fight for basic rights, but things have become much worse for M.E. patients and they continue to grow worse still as the years pass.

M.E. is a shocking disease in every way.

M.E. is at least as disabling as any of the other very serious diseases (such as M.S.) and the extremely high level of suffering and isolation it causes can last for many years or decades at a time. Yet M.E. patients get the least amount of support and compassion and such high levels of abuse and outright *ridicule*.

Some of us have some family and/or friends on board, some have welfare, some have basic medical care (although almost none have the same level of care the average M.S. patient has). But most don't have all or even most of these things and when they do they have often taken many years to get and are very hard won.

By the time many of us have some of these things we have been made severely ill by going so long without the right care, that it's a somewhat hollow victory. Especially when we also know that so many others aren't so lucky and that every year thousands of patients, adults as well as teenagers and very young children, are still needlessly being made severely ill or dead through ignorance and misinformation.

It's like an episode of 'The Twilight Zone.' You want to wake up and scream some mornings, thinking it's a nightmare and that such a hell just couldn't possibly be real. That so many innocent people could be so ill, abused and persecuted, with almost none of the public even caring or knowing. That such a flimsy and unethical global medical scam couldn't be so successful at fooling almost everyone, despite the fact it's based on nothing more than smoke and mirrors, scientifically speaking. It's all just far too shocking to take in sometimes.

I invite readers to be shocked about what is happening, even if M.E. hasn't yet affected someone you love or know. The facts are profoundly shocking – I haven't explained even half of them here.

If you have the facts about M.E. you should be not only shocked by what is happening, but also appalled, disgusted and outraged. I beg you to please use that shock, act on it and use it to help try and change things, and to see M.E. patients finally get some basic fair treatment and justice.

The only way change will occur is through education, with enough people simply refusing to accept what is happening anymore.

M.E. patients need your help so desperately, right now. Thank you for taking the time to read this paper.

To read a fully-referenced version of the medical information in this text compiled using information from the world's leading M.E. experts, please see the 'What is M.E.?' paper on page 113 of this book or on the HFME website.

Acknowledgments
Thanks to Emma Searle, Lesley Ben and Peter Bassett for editing this paper.

Relevant quotes
'People in positions of power are misusing that power against sick people and are using it to further their own vested interests. No-one in authority is listening, at least not until they themselves or their own family

join the ranks of the persecuted, when they too come up against a wall of utter indifference.'
PROFESSOR M. HOOPER 2003

'The term myalgic encephalomyelitis has been included by the World Health Organisation (WHO) in their International Classification of Diseases (ICD), since 1969. It cannot be emphasised too strongly that this recognition emerged from meticulous clinical observation and examination.'
PROFESSOR MALCOLM HOOPER 2006

'Myalgic Encephalomyelitis (M.E.) is distinguished by a unique clinical and epidemiological pattern characteristic of enteroviral infection. It has an UNIQUE neuro-hormonal profile.'
DR ELIZABETH DOWSETT

'The degree of physical incapacity varies greatly, but the [level of severity] is directly related to the length of time the patient persists in physical effort after its onset; put in another way, those patients who are given a period of enforced rest from the onset have the best prognosis.'
DR MELVIN RAMSAY (ON MYALGIC ENCEPHALOMYELITIS)

'Modern medicine is not scientific, it is full of prejudice, illogic and susceptible to advertising. Doctors are not taught to reason, they are programmed to believe in whatever their medical schools teach them and the leading doctors tell them. Over the past 20 years the drug companies have taken medicine over and now control its research, what is taught and the information released to the public.'
ABRAM HOFFER MD

'Central nervous system dysfunction, and in particular, inconsistent CNS dysfunction is, undoubtedly both the chief cause of disability in M.E. and the most critical in the definition of the entire disease process. Of the CNS dysfunctions, cognitive dysfunction is one of the most disabling characteristics of ME. When this simple fact is understood, it become immediately apparent why this is such a devastating disease for children, students and adults, both within and outside the educational system. Today, few work situations exist where consistent use of education and developed cognitive skills are not necessary to maintain a place in the work force. When the patient consistently has difficulty in making new memories, recalling old memories and coordinating new and old information he becomes of little use in the modern work force. It is the combination of the chronicity, the dysfunctions, and the instability, the lack of dependability of these functions that creates 'the most chronic of chronic disabilities.' It is these combined acquired, chronic brain and physical dysfunctions that define M.E.'
DR BYRON HYDE AND ANIL JAIN MD IN 'THE CLINICAL AND SCIENTIFIC BASIS OF M.E. P 43

'Suppose that, in the 'bad old days' before polio vaccination, a parent whose child had died had been told: "She stopped breathing on purpose you know." A public outcry would surely have ensued. And imagine if the next remark had been: "Tell me, did you encourage her in this belief that she couldn't breathe?" The mere idea of such an attitude, quite properly, takes the breath away. Yet children with severe M.E., unable to walk or even to eat, are often considered to be shamming and all sorts of bizarre strategies have been used to try to expose this.'
JANE COLBY IN 'M.E.: THE NEW PLAGUE' P.22

'The vested interests of the Insurance companies and their advisers must be totally removed from all aspects of benefit assessments. There must be a proper recognition that these subverted processes have worked greatly to the disadvantage of people suffering from a major organic illness that requires essential support of which the easiest to provide is financial. The poverty and isolation to which many people have been reduced by M.E. is a scandal and obscenity.'
PROFESSOR MALCOLM HOOPER 2006

What is M.E.? Summary

COPYRIGHT © JODI BASSETT 2004. UPDATED JUNE 2011. FROM WWW.HFME.ORG

 Myalgic Encephalomyelitis (M.E.) is a debilitating acquired neurological disease that has been recognised by the World Health Organisation (WHO) since 1969 as a distinct organic neurological disorder.

M.E. can occur in both epidemic and sporadic forms; over 60 outbreaks of M.E. have been recorded worldwide since 1934.

M.E. is similar in a number of significant ways to Multiple Sclerosis, Lupus and Poliomyelitis (Polio). It can become extremely severe and disabling and in some cases is fatal.

Is M.E. a new illness?

No. The illness has been documented as an organic (physical) neurological disease for centuries.

The name Myalgic Encephalomyelitis was coined in 1956 in the UK.

M.E. has nothing to do with 'fatigue'

Unlike 'Chronic Fatigue Syndrome' (CFS) M.E. is a neurological illness of extraordinarily incapacitating dimensions that affects virtually every bodily system. Fatigue is not a defining (or essential) symptom of M.E. M.E. and 'CFS' are not at all the same thing.

Why do some groups claim that M.E. and 'CFS' are synonymous terms?

This new name and case definition of 'CFS' was created in the United States by a board of 18 members, few of which had either looked at an epidemic of M.E. or examined *any* patients with the illness.

Why? Money! In the late 1970s and 1980s there was an enormous rise in the reported incidence of M.E. causing alarm among American medical insurance companies.

It was at this time when, in order to side-step the financial responsibility of the many new incoming claims, those involved in the medical insurance industry (on both sides of the Atlantic) began their campaign to reclassify this severely incapacitating and discrete neurological illness as a psychological or 'personality' disorder.

As Professor Malcolm Hooper explains:

> A political decision was taken to rename M.E. as "CFS", the cardinal feature of which was to be chronic or on going "fatigue", a symptom so universal that any insurance claim based on "tiredness" could be expediently denied. The new case definition bore little relation to M.E.: objections were raised by experienced international clinicians, but all objections were ignored.

Public, medical and governmental understanding of M.E. is a huge mess, that is for certain – but it is not an *accidental* mess. (For more information see: <u>Who benefits from 'CFS' and 'ME/CFS'?</u>)

What does a diagnosis of 'CFS' actually mean?

Those diagnosed using the flawed 'CFS' definitions are from a heterogeneous (mixed) population with various misdiagnosed psychiatric and miscellaneous non-psychiatric states that have little in common except the symptom of fatigue. 'CFS' is a wastebasket diagnosis; a mere diagnosis of exclusion.

The fact that a person qualifies for a diagnosis of 'CFS' based on any of the 'CFS' definitions (a) does not mean the patient has M.E., and (b) does not mean she or he has any other distinct and specific illness named 'CFS.' A diagnosis of 'CFS' – based on any of the 'CFS' definitions – can only ever be a *mis*diagnosis.

What is M.E.? What is its symptomatology?

M.E. is characterised primarily by damage to the central nervous system (the brain) initiated by an enteroviral infection that results in dysfunctions and damage to many of the body's vital systems as well as a loss of normal internal homeostasis.

M.E. symptoms are manifested by virtually all bodily systems including: cognitive, cardiac, cardiovascular, immunological, endocrinological, respiratory, hormonal, gastrointestinal and musculo-skeletal dysfunctions and damage. These symptoms are exacerbated by physical and cognitive activity, sensory input and orthostatic stress beyond the individual's limits. In addition to the risk of relapse, repeated or severe overexertion can also cause permanent damage (e.g. to the heart), disease progression and/or death. Symptoms of M.E. include:

> Sore throat, chills, sweats, low body temperature, low grade fever, lymphadenopathy, muscle weakness (or paralysis), muscle pain, muscle twitches or spasms, hair loss, nausea, vomiting, vertigo, cardiac arrhythmia, orthostatic tachycardia, orthostatic fainting or faintness, photophobia and other visual and neurological disturbances, hyperacusis, alcohol intolerance, gastrointestinal and digestive disturbances, allergies and sensitivities to many previously well-tolerated foods, drug sensitivities, stroke-like episodes, nystagmus, difficulty swallowing, myoclonus, temporal lobe and other types of seizures, an inability to maintain consciousness for more than short periods at a time breathing difficulties, emotional lability and sleep disorders.

> Cognitive dysfunction may be pronounced and can include: difficulty/loss of ability in speaking or understanding speech; difficulty in reading, writing or performing basic mathematical tasks as well as having problems with memory including difficulty making new memories and recalling formed memories; difficulties with visual and verbal recall.

What does cause M.E.? Are there outbreaks?

A review of early outbreaks in the history of M.E. shows clinical symptoms were consistent in over 60 recorded epidemics spread all over the world as far back as 1934. M.E. is an acutely acquired neurological illness initiated by a viral (enteroviral) infection with a 4-7 day incubation period. This point of view is supported by history, incidence, symptoms and similarities with other viral illnesses as well as a large body of research spanning decades.

So what do we know about M.E. so far?

There is an abundance of research that shows M.E. is an organic illness that can have profound effects on many bodily systems. Many aspects of the pathophysiology of the disease have been medically explained, and to date there are volumes of articles written, from which more than a thousand good articles support the basic premise of M.E. While there is yet no *single* laboratory test able to diagnose M.E., there *are* a specific *series of tests* which enable an M.E. diagnosis to be easily confirmed; i.e. MRI and SPECT scans of the brain.

Some of the abnormalities found in M.E. patients include: extremely low circulating blood volume (up to an astounding 50%), enzyme pathway disruptions, punctate lesions in M.E. brains resembling those of Multiple

Sclerosis; sub-optimal cardiac function and abnormal cardiovascular responses; persistent viral infection in the heart, severe mitochondrial defects and significantly reduced lung functioning.

Also, strong evidence exists to show (even mild or moderate) exercise can have extremely harmful effects on M.E. patients; permanent damage may be caused as well as disease progression and even death. For this reason, danger exists when medical professionals recommend (and sometimes insist on or even *force*) M.E. patients, including children, to partake in exercise as a treatment to their diagnosis of 'CFS.' Under these harmful circumstances, the M.E. patient is undergoing what amounts to actual legalised torture. Patient accounts of exiting exercise programs much more severely ill than when they entered them, *being wheelchair-bound, bed-bound or needing intensive care* are common. *Deaths have also been reported in M.E. patients following exercise.*

How common is M.E. and who gets it?

M.E. has a similar strike rate to Multiple Sclerosis. M.E. affects more than one million children as young as five, as well as teenagers and adults. It affects all ethnic and socio-economic groups, and has been diagnosed all over the world.

Recovery from and severity of M.E.

M.E. can be progressive, degenerative (change of tissue to a lower or less functioning form, as in heart failure), chronic, or relapsing and remitting. It can also be fatal. Patients who are given advice to rest in the early stages of the illness (and who avoid overexertion thereafter) have repeatedly been shown to have the most positive long-term prognosis. M.E. is a life-long disability where relapse is always possible. Symptoms are extremely severe for at least 30% of sufferers leaving many of them housebound, bedbound and severely disabled.

Truly M.E. can be one of the most devastating and horrific illness there is, yet many with M.E. are subject to repeated medical abuse and neglect because of the way the illness has been dishonestly 'marketed' to the public as being psychological or 'behavioural,' or as being a problem of mere 'fatigue' or a 'fatigue syndrome.'

Sub-grouping or refining or renaming 'CFS' will only waste another 20 years. *There is no such distinct disease/s as 'CFS.'* For the benefit of all the patient groups involved, the bogus disease category of 'CFS' must be abandoned and patients with M.E. must again be diagnosed with M.E. and treated for M.E.

Due to an overwhelming amount of compelling scientific evidence, in 1969 the World Health Organization correctly classified M.E. as a distinct organic neurological disease. This classification/definition and name must be accepted and adhered to in all official documentations and government policy.

PLEASE help to spread the truth about Myalgic Encephalomyelitis.

This appalling abuse and neglect of so many severely ill and vulnerable people on such an industrial scale is inhumane and has already gone on far too long. This will only change through education.

People with M.E. desperately need your help.

To read a fully-referenced version of the medical information in this text compiled using information from the world's leading M.E. experts, please see the "What is M.E.?" paper on page 113 of this book or on the HFME website.

Acknowledgments
Thanks to Roseanne Schoof and Emma Searle for editing this paper.

Relevant quotes

'Do not for one minute believe that CFS is simply another name for Myalgic Encephalomyelitis (M.E.). It is not. The CDC definition is not a disease process.'
DR BYRON HYDE 2006

'Myalgic Encephalomyelitis is a clearly defined disease process. CFS by definition has always been a syndrome. M.E. and CFS are not the same.'
DR BYRON HYDE 2006

'Thirty years ago when a patient presented to a hospital clinic with unexplained fatigue, any medical school physician would search for an occult malignancy, cardiac or other organ disease, or chronic infection. The concept that there is an entity called chronic fatigue syndrome has totally altered that essential medical guideline. Patients are now being diagnosed with CFS as though it were a disease. It is not. It is a patchwork of symptoms that could mean anything.'
DR BYRON HYDE 2003

In the mid 1980s, the incidence of M.E. had increased by some seven times in Canada and the UK, while in the USA a major outbreak at Lake Tahoe (wrongly ascribed at first to a herpes virus) led to calls for a new name and new definition for the disease, more descriptive of herpes infection. This definition based on "fatigue" (a symptom common to hundreds of diseases and to normal life, but not a distinguishing feature of myalgic encephalomyelitis) was designed to facilitate research funded by the manufacturers of new anti-herpes drugs. However, a "fatigue" definition (which also omits any reference to children) has proved disastrous for research in the current decade.
RESEARCH INTO M.E. 1988 - 1998 TOO MUCH PHILOSOPHY AND TOO LITTLE BASIC SCIENCE BY DR ELIZABETH DOWSETT

'Fatigue is immeasurable and largely indefinable. Fatigue is a normal phenomenon as well as being associated with almost all chronic disease states. Fatigue, which is simply one of the common features of healthy life and disease, neither defines M.E. nor clarifies the illness. The term 'fatigue' does cause disparagement to those who study this serious debilitating illness and those who suffer from it.'
DR BYRON HYDE

'There are actually 30 well documented causes of 'chronic fatigue'. To say that M.E. is a 'subset' of CFS is just as ridiculous as to say it is a 'subset' of diabetes or Japanese B encephalitis or one of the manifestly absurd psychiatric diagnosis, such as, 'personality disorder' or 'somatisation.''
DR ELIZABETH DOWSETT

'The human rights of people suffering from M.E. are being conspicuously denied without any justification whatever. It is completely unacceptable that the unsubstantiated personal beliefs of a few immensely influential psychiatrists with indisputable vested interests should continue to indoctrinate UK medicine and the media regarding M.E. and that these psychiatrists should be permitted to impose inappropriate management regimes upon sick and defenceless patients on pain of having their benefits withdrawn if they do not comply, a situation that has continued unabated for far too long.'
PROFESSOR M. HOOPER AND E.P. MARSHALL IN M.E.: WHY NO ACCOUNTABILITY?

'M.E. [is] a loss of the ability of the central nervous system (CNS) to adequately receive, interpret, store and recover information. This dysfunction also results in the inability of the CNS to consistently programme and achieve normal smooth end organ response. [It is a] loss of normal internal homeostasis. The neurochemical homeostatic events continue to be employed uselessly and to the detriment of the organism. This modulatory biochemical complex, biologically derived over the millennium to assist the organism, destabilises the autonomic neuronal outflow and the individual can no longer function systemically within normal limits.'
DR BYRON HYDE

CHAPTER TWO
Information for friends, family and partners

This chapter includes the following papers:

1. So you know someone with M.E.?

If you know someone with Myalgic Encephalomyelitis (M.E.) and want to know how to deal with it, and what you can do to help, then this paper is for you.

2. So you know someone with M.E.? Part 2: Tips on coping for friends, partners and family members

If someone close to you has M.E. and you'd like some tips on how to cope and what this might mean for you, and perhaps also your family, then this paper has been written for you.

3. The misdiagnosis letter for M.E. patients

This letter contains blank spaces and is designed to be *filled out and personalised* by the person with M.E. that has loaned or given you this book – if they desire to do so. (It should be used only by M.E. patients.) This letter explains that the patient's correct diagnosis is in fact M.E.

If this book was not loaned or given to you by a specific M.E. patient, then this section of the book may be skipped. (If you aren't sure if this section has been filled in for you to read, do make sure to check.)

If writing in the book directly is not advisable, carers or patients can also either download a copy of this letter (free) from the HFME website, or fill in a photocopy of the appropriate pages from this book.

So you know someone with M.E.?

So you know someone with Myalgic Encephalomyelitis (M.E.) and would like to educate yourself in order to understand it and be of help? Good on you! Here are some suggestions on where to start:

(For the purpose of avoiding wordy repetition, from this point forward, *any person in your life who suffers from M.E. – a friend, partner, family member, co-worker, neighbour, etc. – will be referred to simply as 'friend.' Also, for the sake of continuity, the female pronoun will be used throughout.*)

1. Offer acceptance, respect and emotional support

Due to the financially and politically motivated propaganda on the subject, many people with M.E. have received horrible treatment and abuse from friends and family as well as the medical profession. This cruel propaganda was actually planned in order to create an intentional 'confusion' between M.E. and 'CFS' in the eyes of the public and even the very people whose lives M.E. devastated.

Because M.E. is a serious neurological disease, you need to treat your friend no differently than you would treat her if she had Multiple Sclerosis or any other politically accepted disease. Medically speaking, M.E. is *very* similar to M.S. It shares medical similarities with Polio and Lupus as well. Treat your friend the way *you* would want to be treated if *you* had contracted a horrible disease. M.E. is hell, but without the support of family and friends it can be indescribably worse.

Your friend faces the great possibility of serious medical abuse and/or neglect, outright disbelief, and even cruel ridicule and isolation as she begins to seek help from doctors and peers. This is where even basic kindness and respect on your part should not be underestimated. Your support can make all the difference in the world.

2. Learn the basic facts about M.E.

Learning more about M.E. will enable you to act more appropriately towards your suffering friend. It will help you in communicating with her as she opens up to you and shares her experience; and you will be a qualified advocate for her when the time comes for you to educate others about this disease.

Know that M.E. is not:

- Fatigue or 'CFS' or 'CFIDS' or 'ME/CFS'
- Medically unexplained or mysterious. See *What is M.E.?*

Know that M.E. *IS:*

- A serious neurological disease that is similar in many ways to M.S. and has more than 60 neurological, cardiac, metabolic, gastrointestinal, immunological and other symptoms.
- A disease that occurs in epidemic and sporadic forms and *can be tested* by using a series of objective tests (including MRI and SPECT brain scans). See *M.E. vs. M.S.: Similarities and differences.*

- A disease where physical activity beyond the patient's limits makes her much sicker, even when it is seemingly *minor*; including being in an upright position, receiving sensory input such as light or sound, or even thinking.
- A disease where relapses can be severe and can last for hours or as long as days, weeks, months or even years; relapse can even be permanent. (Death can also occur with severe or *repeated* overexertion in M.E.) See: *The importance of avoiding overexertion in M.E.* on the HFME site for more information.

The quickest and best way to get some of the basic facts about M.E. is to read *What is M.E.?* or the shorter version of this paper: *What is M.E.? Summary*. Other highly recommended reading is: *A Million Stories Untold* - written specifically for family and friends of M.E. patients.

It is important to be aware of the vast amount of incredibly harmful and unreliable information you will inevitably encounter on this topic. The mainstream media and government are not credible information sources; nor are almost all so-called advocacy groups which claim to advocate for M.E. or 'CFS', 'CFIDS' or 'ME/CFS' patients. Be aware that many of these unhelpful groups dare to use the term M.E. in some way to peddle their 'CFS' and 'ME/CFS' misinformation, whether in the name of their group or in its literature.

There are large advocacy groups throughout the different countries that have clearly sold out patients for the groups' own benefits.

To read more about this problem, and to learn the names of the few groups that *are* reputable, see: *Problems with 'our' M.E. (or 'CFS' 'CFIDS' or 'ME/CFS' etc.) advocacy groups*. Also recommended is *My comments about the current (worrying) state of Australian 'CFS/ME' societies* and *Problems with the so-called "Fair name" campaign: Why it is in the best interests of all patient groups involved to reject and strongly oppose this misleading and counter-productive proposal to rename 'CFS'*

For more understanding on the why and how of this abuse, see: *Who benefits from 'CFS' and 'ME/CFS'?* Having M.E. is hell, but having M.E. and being treated unfairly or inappropriately by friends and family is beyond description

An additional note on M.E. education: It is also recommended that you learn how M.E. affects your *particular* friend. While M.E. is very consistent from one individual to the next in its major features, each patient's experience of it is individual to some extent; the severity of different symptoms varies from patient to patient. If you don't know how the disease affects your particular friend, *ask her*. She will appreciate your desire to understand and help. Being specific in your questions will not offend her. Making futile attempts at being helpful can cause stress for your friend, possibly causing a relapse of her symptoms. Even small overexertions can start an entire cycle of problems for your friend, ending with her body breaking down and needing a great deal of rest.

As obvious as this might seem to be, it is important to instruct you to **believe** your friend when she describes her symptoms, experience and severity. M.E. patients typically *do not* exaggerate or lie about their symptoms or their disability level, yet they will often encounter alienating and demoralising scepticism or disbelief from doctors and even friends and family.

The simple acceptance on your part goes a long way in showing your respect, and will mean a lot to your suffering friend.

Because your friend does not wish to complain *again*, and appear to be a 'whinger' or a 'whiner,' you might find she has been tolerating excruciating pain from your voice and choices of volume on the TV or radio, for example. For you, the volume is at a normal balance; for your friend, it could be causing pain in her ears, minor seizures and make it impossible for her to focus or to think. She may be suffering a great deal of pain not only during the time listening, but also later, alone in her room, where the neurological pain and ear pain and perhaps even seizures will keep her up for several nights or longer and perhaps even having sent her into a long-time relapse with permanent damage.

Therefore, adapting your behaviour to accommodate particular symptoms is more meaningful than you can ever fully understand. The small gesture will save your friend a great deal of suffering, and it will save a friendship as well.

A friend who understands these seemingly small things is a very precious thing to her M.E. friend.

3. Help the person with M.E. to rest, to avoid the overexertion which can lead to disease progression

Rest is extremely important for the M.E. patient, but sometimes they can be so desperate for conversation and social connection and just plain old FUN, that it can be difficult to moderate (or deny themselves) this rare gift. You can help by:

- Being adaptable about communication modes. If someone tells you she can't talk on the phone anymore but is able to have short visits, or can *only* talk via email, please accommodate her as much as you can.
- Asking beforehand about the best time of day for calls or visits.
- *Planning no surprise visits* EVER! You may put your friend in a very awkward position. She may either have to risk being rude by asking you to leave, or something that happens in most cases will be that she will not feel free to speak openly, causing her to suffer through the visit, whereas she could have thoroughly enjoyed a pre-planned, timed visit.
- Setting a time limit beforehand for calls or visits and making it easy for her to stick to it.
- Should your friend suddenly look unwell during your visit, asking her outright if she needs to be left alone will take the burden off her shoulders of trying to find a polite way to take the initiative.
- Being gracious when someone with M.E. has to cancel a visit or a call; these cancellations will take place last minute most of the time; please be understanding. Because this disease gives no notice of a flare, cancelling is something she is doing often, and it causes her to feel a lot of guilt.
- Being gracious when someone with M.E. has to suddenly or prematurely end a visit or call. With M.E., a patient can be coping well during an activity or conversation, or be enjoying a TV program when, without notice she will become very ill, experiencing a variety of severe neurological, cardiac and/or other symptoms, forcing her to need to stop immediately and retreat to a dark, isolated, quiet place.
- Not pressuring a suffering friend to perform an activity or participate in a conversation after she has already made it clear she cannot do so. Often, friends will feel they can gently 'force' the patient to do something because it will make her 'feel better'…remember, you have no idea how she suffers.
- Sometimes, friends of M.E. patients feel rejected or ignored when the friend has not responded to e-mail, phone calls or invitations. Try to remember; no matter how many times this happens it is *not personal*. If this bothers *you, try to imagine how it bothers her*. She *does* desire to spend time with you. Not being able to respond and not being able to be with you is difficult. Most times, your friend is lonely and desperate for company yet she cannot do anything about it because of this awful disease. Before you get angry, please remember this.
- When a person is very severely ill, the only contact she may be able to tolerate is for you to sit with her for a little while, without talking, in a darkened room; perhaps gently taking her hand in yours or placing a hand on a leg or arm is possible without much discomfort. Human contact is very important, but especially to people who do not get it very much, and these patients fall into that category.

It would also be great if you could not tempt M.E. patients to deviate from their often, very strict diets. Pressuring her to take a sip of your alcoholic or sugared drink or to take a bite of a delicious dish you made puts her in an uncomfortable and illness-worsening place of having to turn you down and explaining why. Try to not place your friend in this kind of situation; allow her to use her 'extra cardiac output' purely on enjoying your company.

Remember, having so many restrictions and limitations is neither fun nor easy. They are difficult to stick to, and when she gives in to temptation and takes that sip or bite of food, your friend suffers from the negative effects usually long after you have gone home.

M.E. is a devastating disease, but the good news is that you really do have the power to make an enormous positive difference to your suffering friend – just by being there for her and offering practical and/or emotional support and also by helping her avoid overexertion as much as possible and so having her best possible prognosis.

You do not need to find articles for her to read; you do not need to find something exciting for her to do or taste or think about.

All she needs from you is YOU. Your company.

4. Encouraging your M.E. friend to be more active can harm her

Even trivial levels of activity over her individual post-illness limits can cause severe relapse, as much as leaving your friend wheelchair dependant or bed-bound for many long days, weeks and even *years* afterward, even permanently. Overexertion can cause unrecoverable cardiac insufficiency and/or congestive heart failure in M.E. which may result in death. You do not want to contribute to such a tragic outcome.

Increasing activity levels is something every person with M.E. will do the second she is able. But arbitrarily increasing a person's activity levels above what she can cope with can only *ever* be *counterproductive*. It can be compared to insisting someone with two broken legs take up jogging; it is extremely painful, damaging and cruel – and of no possible benefit.

5. Remember that looks are deceiving; just because she looks well does not mean that she is doing well

Try not to make superficial judgements of ability or severity! An M.E. patient's appearance does not indicate the severity of her illness. While many M.E. patients do look ill, many M.E. patients do not look as ill as they are. Furthermore, an M.E. patient's activity over a particular period does not indicate her abilities over the long term.

In *most* illnesses, one might be able to observe a patient carefully for short periods of time and successfully collect a good amount of useful information regarding the patient's abilities and restrictions i.e. the severity of her illness, average daily symptoms, etc.; however M.E. *is not* such an illness. M.E. is *not* a stable illness.

Observing the average M.E. sufferer for an hour – or even a week or more – will *not* give an accurate indication of her usual activity level because the severity of M.E. can wax and wane throughout the day, week, month, or even down to the very hour. Also, people with M.E. can sometimes operate *significantly above* their actual illness level for short periods of time due to surges of adrenaline – albeit at the cost of severe and prolonged worsening of the illness afterward. Relapses and worsening of symptoms are also very often significantly delayed (there will most often be both an acute *and* a delayed reaction).

Note that M.E. patients are often observed functioning at a level which is not at all representative of their illness. For example, when they have visitors or when they are in public, they will often push themselves to perform at a higher level than they ought, thereby causing a delayed relapse which usually occurs in private, outside the observation of those who had seen them earlier that day or that week.

Simple observation of someone with M.E. performing a certain task should *not* be taken to mean that: (a) she can necessarily repeat the task at any future hour, day or even week or month (b) she would have been able to do it at any other time of day (c) she is also able to do all (or any) seeming similar tasks (d) the patient could have done this same task without a rest period beforehand of days or weeks (e) the patient won't be made very ill afterwards for a considerable period of time. Performing the task may have taken every bit of any strength she might have had in reserve, causing relapse and possibly disease progression.

Often, a considerable rest period of hours, days, weeks or months is needed before and after performing a task. For example, someone may need 2 weeks rest before an outing and may then spend 3 weeks extremely ill recovering from it afterwards. Just observing her in the 2 hours she was 'out and about and mobile' is of course not at all representative of her usual ability levels.

Most importantly, because the worsening of the illness caused by overexertion may not even begin until 48 or more hours afterwards (when most observers are long gone) it's impossible to tell by seeing an M.E. patient engaged in an activity whether that activity is so far beyond the patient's limits that it will end up causing a severe or even permanent worsening of the illness (or 'relapse'). To be blunt, although uncommon (3% of the time), such activity can even end up *killing* the patient. Observers who see an M.E. patient engaged in an activity have no idea what the consequences of this activity may be.

One simply cannot know an M.E. sufferer's usual ability level or severity level except by observing it over a very long period of time, or actually **asking** the patient detailed questions about her average daily activity limits, abilities and symptoms. Short-term and superficial judgements of ability and disability levels in people with M.E. are ill-advised. In people with M.E., appearances are *almost always* very deceiving.

See *The M.E. symptom list* for more information, plus *Why patients with severe M.E. are housebound and bedbound* and *Hospital or carer notes for M.E.*

6. The latest and greatest advice being doled out to patients of all illnesses is to think positive; patients are to 'think themselves well'. In any patient, this is wrong, but in cases with M.E., it is not helpful or kind or reasonable. Do not ask your friend to consider such nonsense

There is no more possibility that M.E. could be improved or cured through positive thinking or by will power any more than M.S., Parkinson's, or a broken leg! Personality types and attitudes have nothing to do with the prognosis of M.E. any more than with other illnesses.

Studies showing positive outcomes for exercise and positive thinking (GET and CBT) on tired people are irrelevant to people with M.E., or those with any other distinct neurological disease.

Telling someone with a serious organic, neurological disease that she is ill only because she 'thinks' she is ill or that she could 'think' herself well if she simply tried hard enough is not only abusive and incredibly cruel, but can be *very harmful*. As stated above, when M.E. patients are influenced to push themselves beyond their physical limits, they can seriously damage their health (even unto congestive heart failure or death in some cases).

For more information see: *The effects of CBT and GET on patients with M.E.* and *Comments on the 'Lightning Process' (etc.) scam* on the HFME website.

7. Please don't recommend treatments you may have read about online, or in the paper, or heard about from friends

Very close to 100% of media articles which purport to be about M.E. are in fact talking about various subgroups of chronic fatigue or *'CFS'* – people with all sorts of very different and often much milder and/or transient diseases.

There is always the *'friend's brother's neighbour's second cousin twice removed'* who supposedly had identical symptoms as your friend, and who 'recovered completely' and is now 'back working full time' all because of treatment "x" mentioned in a particular article.

Bogus treatment recommendations can leave an M.E. patient far more severely ill and broke. The biggest 'cost' from being taken in by misleading claims about treatments is not usually the financial cost however, but the high emotional price of false hope. Many M.E. patients have already been wearied and traumatised by having their hopes repeatedly raised and dashed. It is necessary to point out that due to the appalling lack of funding for legitimate research, there is no cure yet in sight, nor are there treatments which can dramatically influence the natural course of the illness once it has been made severe.

Nutritional, pharmaceutical and other developments can make a significant difference to a patient's life. However, the person you know with M.E. will have far more knowledge than you do about where to find legitimate information and which sources to avoid. People with M.E. are extremely well-motivated to get well; if a simple cure for M.E. existed, they would know. This is a fact that can be counted on.

If you absolutely must recommend something, please read *Treating M.E. - The basics* first to make sure the treatment is not already well-known, or known not to be appropriate for M.E. If you are still sure that what you have is 'real', print the information out and send it by post. Also, let the person know that you are not putting any pressure on her to try this medication, assuring her if she does not choose to try it, she owes you no explanation.

Do not offer medical advice unless asked, and even then, be very careful you are not doing more harm than good. And remember, your friend has most likely received many pieces of advice already, and feeling pressured to try another puts a physical and emotional burden on her.

Having M.E. brings many pieces of advice, recommendations and plain arrogance from others; this alone can make enjoying and resting in a friendship very difficult for the patient.

8. Some M.E. patients may appreciate some practical help

You might like to ask your friend with M.E. if she is in this situation, and then you may then offer to:

- Do some shopping for her on a regular basis
- Help with meals or other household tasks
- Drive her to medical appointments and/or book medical or other appointments
- Research something for her online, e.g. disability services
- Do anything she might need doing that you are able to do

If you have any involvement in the care of the person with M.E., you may like to read:

- Hospital or carer notes for M.E.
- Why patients with severe M.E. are housebound and bedbound
- The importance of avoiding overexertion in M.E.
- Treating M.E. - The basics
- The myths about M.E.

9. Be a good friend and allow her continue to be a friend to you

Although your friend is dealing with severe illness, she does not wish to exclude your daily struggles or problems. She might need you to lean on for support and understanding, but as in any friendship, she not only wants you to share with her, she *needs* you to do so. She *needs you* to *need her*.

Sometimes it seems to the healthy friend that the conversation gets competitive, or that conversation revolves around the sick friend's problems. Please do not make the mistake of keeping your thoughts and concerns to yourself; share them with your M.E. friend. Sometimes she might be dwelling on her own illness and seemingly forgetting your struggles; speak up and share your life with her. Friendship is important to your M.E. friend now more than ever; having a good friend when a person is housebound and/or bed- bound is a rare gem.

M.E. affects patients physically and changes their experience of life. M.E. may have affected your friend's cognitive function (concentration or word-finding), but try not to be thrown by these symptoms, and remember that your friend is fundamentally the same person he or she always was with the same personality, the same sense of humour, likes, dislikes and quirks.

Your friend may have experienced many important losses, so the one important need your friend has amidst such changes is the continuity of relationships. She may need your friendship now more than ever before. Please cut her some slack if sometimes you find her in a bad mood, or feeling very sad or angry; she still has the same hopes, dreams and desires, and not seeing them realized can be very frustrating and disappointing, especially as she watches you and others moving forward with your lives.

Remember that sounds, lights, any external sources can affect your friend in huge ways; it could seem she is being moody or demanding or even controlling, but even a low-level sound may be affecting your friend in a way you cannot tell. In the case of the M.E. patient, sounds that are not even actually consciously 'heard' by you could be torture for your friend, causing seizures that you are not able to discern.

Although your friend may seem to be happy and coping – and even if they are most of the time – they're still living every day under extreme stress, with extreme pain and suffering with no end in sight. Many of them hide the suffering very well, particularly after many years of being very ill, but that doesn't mean it does not exist. It also does not mean the patient is not going through great feats to simply keep up.

People with M.E. have spent years living in pain and have become quite adept at hiding it; they continue to smile when they see a friendly face; instead of complaining about every ache and pain, they become selective so that when they do make their misery evident, they might be heard and receive some serious help. Do not assume your friend is having a good day. A good day for an M.E. patient is not the same as a good day for you; her good day might mean she was able to get up from her bed or wheelchair for 10 minutes. Her good day might mean she was able to even get outdoors for a short while. It never means she is pain-free. It won't mean her heart has not given her trouble or her vision is no longer doubled or blurred.

Remember that sounding happy is not the same as sounding healthy. Feel free to comment on how happy someone sounds but please don't assume that this means that they are doing well health-wise.

10. Help get other friends and family members informed or better informed

Due to the large amounts of misinformation available on this topic, becoming an educator to your friend's family and others could prove to be an important role you undertake. You might consider explaining the basics of M.E. by sharing a printout of *What is M.E.? Summary.* This information may be received far better if it comes from you, rather than from the patient.

11. Help get the wider community better educated and informed

Distributing the printout mentioned above to friends, neighbours and co-workers and even medical personnel in your area could greatly benefit M.E. patients, as could linking to the HFME site on sites which carry incorrect information on the disease. Consider joining the HFME discussion group or make a donation to help fund M.E. advocacy; encourage others to do the same, see the Donations page on the website.

More information

- Highly recommended is: A Million Stories Untold - this paper was written specifically for family and friends of M.E. patients.
- In short:

    Chronic Fatigue Syndrome is an artificial construct created in the US in 1988 for the benefit of various political and financial vested interest groups. It is a mere diagnosis of exclusion (or wastebasket diagnosis) based on the presence of gradual or acute onset fatigue lasting at least 6 months. If tests show serious abnormalities, a person no longer qualifies for the diagnosis, as 'CFS' is 'medically unexplained.'

    A diagnosis of 'CFS' does not mean that a person has any distinct disease (including M.E.). The patient population diagnosed with 'CFS' is made up of people with a vast array of unrelated illnesses, or with no detectable illness. According to the latest CDC estimates, 2.54% of the population qualifies for a 'CFS' (mis)diagnosis. Every diagnosis of 'CFS' can only ever be a misdiagnosis.

    Myalgic Encephalomyelitis is a systemic neurological disease initiated by a viral infection. M.E. is characterised by (scientifically measurable) damage to the brain, and particularly to the brain stem, resulting in dysfunction and damage to nearly every vital bodily system and a loss of normal internal homeostasis. Substantial evidence indicates that M.E. is caused by an enterovirus. The onset of M.E. is always acute and M.E. can be diagnosed within just a few weeks. M.E. is an easily recognisable, distinct organic neurological disease, which can be verified by objective testing. If all tests are normal, then a diagnosis of M.E. cannot be correct.

M.E. can occur in both epidemic and sporadic forms and can be extremely disabling, even fatal. M.E. is a chronic/lifelong disease that has existed for centuries. It shares similarities with M.S., Lupus and Polio. It has more than 60 different neurological, cognitive, cardiac, metabolic, immunological, and other symptoms. Fatigue is not a defining or even essential symptom of M.E. People with M.E. would give anything to be only severely 'fatigued' instead of having M.E. Far fewer than 0.5% of the population has the distinct neurological disease known since 1956 as Myalgic Encephalomyelitis.

'CFS' is a medical fraud, created (and maintained) for political and financial gain by vested interest groups. People with M.E. are *not* being mistreated because of a lack of scientific evidence; there is an abundance of evidence spanning 70 years, which proves beyond any doubt that M.E. is a distinct, organic neurological disease. Accordingly, Myalgic Encephalomyelitis has been recognised by the *World Health Organisation* (WHO) since 1969 as a distinct, organic neurological disorder. M.E. is classified in the current WHO *International Classification of Diseases* with the neurological code G.93.3.

Medically, M.E. is very similar to diseases such as Multiple Sclerosis. See the *M.E. vs. M.S.: Similarities and differences* paper for details. People with M.E. want to be treated based on the available science and to have their fair share of the resources available to those with medically (if not politically) comparable illnesses such as Multiple Sclerosis or Motor Neurone Disease, etc. Sadly this is not happening, and patients are being mistreated according to political and financial considerations while the science and reality is being ignored. Please help to spread the truth about M.E. and the difference between M.E. and *'CFS.'* Knowledge is power.

To read a fully referenced version of the medical information in this text compiled using information from the world's leading M.E. experts, please see the 'What is M.E.?' paper on page 113 of this book or on the HFME website.

Acknowledgments
Thanks to Roseanne Schoof for editing this paper. Contributors to this paper include Roseanne Schoof, Lesley Ben and Peter Bassett.

Relevant quotes
'At the turn of the millennium, the public still lacks a real grasp on what M.E. patients are dealing with. Because of illusions that M.E. is simply a disease of tired people the public has large been deprived of accurate information.'
LYNN MICHELL IN 'SHATTERED: LIFE WITH M.E.' P XXII

'In all M.E. epidemic or endemic patients the patients represent acute onset illnesses. The fatigue criteria listed [in the 'CFS' definitions] can be found in hundreds of chronic illnesses and clearly defines nothing.'
DR BYRON HYDE 2006

So you know someone with M.E.? Part 2: Tips on coping for friends, partners and family members

COPYRIGHT © JODI BASSETT & LAJLA MARK, FEBRUARY 2010. UPDATED JUNE 2011. FROM WWW.HFME.ORG

 If someone close to you has Myalgic Encephalomyelitis (M.E.) and you would like some tips and useful information on coping, and on helping your friend, partner or family member, then this paper has been created with you in mind.

Here are some suggestions:

1. Know that you did *not* cause the person to contract M.E.
2. Learn the basic facts about M.E. (and about the person's individual illness/limits)
3. Know that it is okay to feel sad and to grieve your own losses
4. Get support from friends and/or others facing the same situation
5. Make good communication between you and the ill person a priority
6. Do not feel guilty for taking time out when necessary
7. Cut yourself some slack
8. Just do your best

(For the purpose of avoiding wordy repetition, from this point forward, *any person in your life who suffers from M.E – a friend, partner, family member, co-worker, neighbour, etc. – will be referred to simply as 'friend'. Also, for the sake of continuity, the female pronoun will be used throughout.*)

1. You did not cause the person to contract M.E.

Know that nothing you did or did not do as a parent caused your child to get M.E. M.E. is caused by a virus which does not discriminate and can affect anyone at any age.

If you in any way feel responsible for having caused another to contract M.E. you must let it go. What matters is what you do now; the right care can make an enormous difference to the prognosis of M.E. and that is something you can affect right now.

2. Learn the basic facts about M.E.

Know that:

- M.E. is not fatigue or 'CFS' or 'CFIDS' or 'ME/CFS' or medically unexplained or mysterious.
- M.E. is a serious and potentially fatal acute-onset neurological disease that is similar in many ways to M.S. and has more than 60 neurological, cardiac, metabolic, immunological and other symptoms. M.E. occurs in epidemic and sporadic forms and can be tested for using a series of objective tests.
- People with M.E. are made much sicker by physical activity (or being upright or even thinking or taking in sensory input such as noise and light) beyond their individual limits. This includes even seemingly minor activities. Relapses can be severe and can last hours, days, weeks, months or even years, or longer.

Be aware that almost all of the information given on this topic in the media and even by charities and government is factually incorrect and grossly misleading.

It is recommended that you learn how M.E. affects the person individually by *asking her* about the illness and where she most needs help and what she needs to avoid. The simple fact of your acceptance of this information shows your respect and may mean a lot to your friend.

It is also very important that you have at least a basic grasp of this information so that you can educate those around you where needed. You will need as much support as possible –*both of you*. (You might like to print up some HFME leaflets or informational business cards for easy redistribution to those around you.)

3. Know that it is okay to feel sad and to grieve your own losses

The person who has M.E. will be going through many different emotions. She will be struggling to accept her changed life and will be in mourning for those losses (freedom, status, job, relationships, and long-held hopes for the future). She will sometimes feel sad, angry, frustrated or completely overwhelmed.

If you are close to someone with M.E. you will be going through many of these same emotions. (Studies have shown that this is the case; close friends and family members can go through the same roller coaster of emotions as the ill person does). Perhaps you miss the other person even though she is still around; you miss the talks and the support she use to give. You need to allow yourself the process of grieving for what you have lost.

4. Get support from friends and/or others facing the same situation

Talking to other people in the same situation can be very helpful; try an online support group. Consulting a psychologist specialised in this area may be another option, provided you can find a person that is understanding and accepting of M.E. as the disabling neurological disease it is. Whatever the choice, it is certain that you will need support.

5. Communication is key

Create good communication between you and your friend and make it a priority; this is not only extremely helpful for the patient, but for the advocate and anyone else involved in the life of the M.E. patient.

Communicating will prevent the patient from being frustrated due to not receiving the right kind of care. This is one case where it is not recommended to do unto others as you would have them do unto you. ASK the patient how she wants a certain task completed. It is not actual help if it is not appropriate to her actual needs, and it is double work if the task has to be done over. Your friend may prefer you to stay with her and talk or do things with her rather than doing the dishes or cleaning the house etc. Or she may prefer the independence of feeding and bathing herself (even if it's very difficult for her physically), rather than having you do it.

You may be able to cut back what you are doing in some areas and perhaps other areas may need slightly more attention or a different type of attention. The only way to know which tasks are the most important for you to do is to ask the patient. Talk to each other about how you can tackle the situation together.

6. Do not feel guilty about taking some time out when necessary

Having time and space to yourself is very important so that you do not become overwhelmed as a caregiver or advocate. Be open with the patient about how and when this can best be done.

If you are the other part of a couple, try to fit a caregiver into your finances. Even if it is a part-time person to clean the house; you want to be careful to prevent a dangerous shift in your relationship where you become the 'parent. It is important to continue your life together as a couple, making sure you keep her dignity in tact. (i.e. Bathing your wife when you were young was sexy; doing so now could be demeaning). Keep the fires burning even if it is a low ember.

Although you might have to make adjustments about the amount of time spent on other parts of your life, keep your favourite hobbies and recreations and friends. You need the space and time off sometimes – and the fun and enjoyment – and this will also give you and your ill partner/family member something else than illness to talk about.

Chances are that even of the ill person can't do many things they would like to do and used to do, they still very much want you to be able to do these things and to enjoy life as much as possible. But do not shut her out; share these things; keep her a part of them. Your M.E. friend needs you, but this must be balanced with your own needs, both practical and emotional.

7. Cut yourself some slack

You are dealing with a difficult situation that is long-term; the illness of your friend is going to take its toll on you as well, causing emotional swings, stress, overwhelmed with the amount of help is being needed from you. You are allowed to have off-days. Cut the person who is ill some slack, as well as yourself. You are both only human. Be kind to yourself. Be your own best friend.

8. Just do your best

There is no perfect way to deal with all of this, so all you can do is your best.

When the patient is a child: As you would spend special time just with the ill child, make sure you set aside special one-one-one time with the healthy children so they don't feel left out and unimportant, or start resenting the ill child for the extra time she receives from her parents

When you are the other half of a relationship: Remember to 'nurse' your relationship with your loved one. She has been your friend for all this time, so make sure you work at keeping that relationship in tact; do not become overprotective by doing too much for her. She will keep her dignity by doing for herself as much as she is able. Do not 'melt' together to a place where you are no longer two separate people; be careful to remain separate; you remain who you are, she remains who she is. Take care but don't take over. Don't forget to laugh.

When you are the other half of an intimate partnership: You have been her husband/wife/partner all these years; what is left of this relationship is the *memory* of what it was *at its best*. Nurture your life together; remember all your happy times together by looking at old photos now and then. Even where physical intimacy is no longer a part of your life, remember the whole person. But even then, continue to show affection, tenderness and love as this can still be very meaningful for the ill person and for you too. You are together because of who you are individually and what you became as one. Don't allow her to doubt your love. She is vulnerable now. This disease has taken its toll on her strength and vitality and perhaps also her physical appearance; she needs reassurance. Create *new* memories and laugh together as much as you both possibly can. Do not become a statistic; fight for the relationship.

Acknowledgments
Thanks to Roseanne Schoof and Emma Searle for editing this paper.

The misdiagnosis letter for M.E. patients

COPYRIGHT © JODI BASSETT, DECEMBER 2010. UPDATED JUNE 2011. FROM WWW.HFME.ORG
This form is designed to be filled out by the M.E. patient (that no longer wishes to be bound by a 'CFS' misdiagnosis) and then shown or sent to friends, family members or carers

Hello _____,

Some time ago I let you know that I was ill. I also told you in the past that the reason I was ill was that I had Chronic Fatigue Syndrome or CFS (or perhaps CFIDS or ME/CFS or CFS/ME).

I would now like to let you know that this was in fact a *mis*diagnosis.

It turns out that the field of 'CFS' is not at all what it first seems. On this topic our medical system, media and even government have been less than forthcoming with the facts. The amount of misinformation circulating is immense. The public, doctors and *even most patients themselves* have been deceived about the reality of 'CFS.'

To make a long story short, there was a huge increase in the rates of a devastating neurological disease called Myalgic Encephalomyelitis (M.E.) in the 1980s in the USA. In response to this, some medical insurance companies (and others) decided that they would prefer not to lose many millions of dollars on so many new claims and so they created a new vague fictional disease category called 'Chronic Fatigue Syndrome' to try to confuse the issue of M.E., and to hide M.E. in plain sight.

They then made the definition of 'CFS' so broad and vague that just about anyone could be diagnosed with it, which saved them even more money as many of these insurance claims could now also be easily denied. To summarise:

A. *Myalgic Encephalomyelitis* is a distinct neurological disease. M.E. is characterised by scientifically measurable damage to the brain caused by a virus. The onset of M.E. is always acute (sudden) and M.E. can be diagnosed within just a few weeks.

M.E. is an easily recognisable distinct organic neurological disease which can be verified by objective testing. If all tests are normal, then a diagnosis of M.E. cannot be correct.

M.E. can occur in both epidemic and sporadic forms and can be extremely disabling, or sometimes fatal. M.E. is a chronic/lifelong disease that has existed for centuries. It shares similarities with M.S., Lupus and Polio. There are more than 60 different neurological, cognitive, cardiac, metabolic, immunological, and other M.E. symptoms. Fatigue is not a defining or even an essential symptom of M.E. People with M.E. would give anything to be only severely 'fatigued' instead of having M.E. Far fewer than 0.5% of the population has the distinct neurological disease known since 1956 as Myalgic Encephalomyelitis.

B. *Chronic Fatigue Syndrome* in contrast is an artificial construct created in the US in 1988 for the benefit of various political and financial vested interest groups. It is a mere diagnosis of exclusion (or wastebasket diagnosis) based on the presence of gradual or acute onset fatigue lasting at least 6 months. If tests show serious abnormalities, a person no longer qualifies for the diagnosis, as 'CFS' is 'medically unexplained.' A diagnosis of 'CFS' does not mean that a person has any distinct disease (including M.E.).

Every diagnosis of 'CFS' can only ever be a misdiagnosis. However, while 'CFS' is not a genuine diagnosis, those given this misdiagnosis are in many cases significantly or even severely ill and disabled, with any number of well over 100 different conditions including post-viral fatigue syndromes, cancer, M.S., Lyme disease, Fibromyalgia, Candida, depression, PTSD and many, many others. All a diagnosis of 'CFS' actually means is that the patient has a gradual onset fatigue syndrome, which is sometimes due to a missed major disease. The patient population diagnosed with 'CFS' is made up of people with a vast array of unrelated illnesses, or with no detectable illness. According to the latest CDC estimates, 2.54% of the population qualifies for a 'CFS' (mis)diagnosis.

Under the cover of 'CFS' certain vested interest groups have assiduously attempted to obliterate recorded medical history of M.E. even though the existing evidence has been published in prestigious peer-reviewed journals around the world and spans over 70 years. The 'CFS' concept has meant that millions of people have been denied correct diagnosis and treatment and have also often been subjected to abuse and a lack of appropriate support.

So now for all the patients out there like me, it's a nightmare.

The only way forward, for the benefit of society and all the different patient groups involved, is that:

1. The bogus disease category of 'CFS' must be abandoned completely, along with all the other similarly vague, misleading and unhelpful umbrella terms such as 'ME/CFS,' 'CFS/ME,' 'ME-CFS,' 'CFIDS,' 'Myalgic Encephalopathy' and others.

2. The name Myalgic Encephalomyelitis must be fully restored (to the exclusion of all others) and the World Health Organization classification of M.E. as a distinct neurological disease must be accepted and adhered to in all official documentations and government policy. M.E. patients must again be diagnosed with M.E. and treated appropriately for M.E. based on actual M.E. research as they were prior to 1988.

3. All those misdiagnosed with 'CFS' must immediately reject this harmful misdiagnosis and begin the search to find their correct diagnosis. Every patient deserves the best possible opportunity for appropriate treatment for their illness and for recovery and this process must begin with a correct diagnosis, if at all possible. A correct diagnosis is half the battle won.

The concept of 'CFS' denies patients their basic rights and subverts good science and ethics. I refuse to play any role in propping up this bent system and concept and so I reject this misdiagnosis. My illness is not fatigue based and is not medically unexplained, new, mysterious or untestable. It never was. The vast majority of those told they have 'CFS' do not have the same disease as I do.

Please understand that this is absolutely not about me wanting a fancier or scarier sounding name for mere fatigue! It is about a severe neurological disease being covered up by a fictional and vague 'fatigue syndrome' by vested interest groups. It is about the fact that research which involves vague mixed patient groups, instead of only people that have the same disease I do, doesn't help anyone. Just as you can't research diabetes by looking at groups of patients that have broken legs, the flu, rashes or headaches – you can't research M.E. by using vague 'CFS' patient groups that may or may not happen to contain a small percentage of actual M.E. patients! To do so is so unscientific it's laughable.

The bottom line, however, is that in essence nothing has changed in what I am telling you. I'm still saying to you that I'm seriously ill, and this was not and is not my fault or something I can wish, positive think or exercise away. My illness and disability is unchanged and is a reality, unfortunately. Things are very difficult for me due to this terrible disease and I could really use your support and friendship. Please treat me no differently than you would if I had M.S., Parkinson's' disease or Polio or Lupus. M.E. is very similar to each of these diseases medically, just not politically.

Not everyone misdiagnosed with 'CFS' has M.E. In fact, *the vast majority do not*. But for me, M.E. is the correct diagnosis. I wish I had known this earlier, but there is so much poor quality information out there as well as slick, misleading and manipulative support for the bogus 'CFS' or 'ME/CFS' constructs. It took me a while to work out where the truth lay. But at least I know now, and now you know too. This is a very important issue to me so thank you so much for reading this.

Best wishes,

_____

CHAPTER THREE
Information for carers and hospital staff

This chapter includes the following papers:

1. Hospital or carer notes for M.E. patients

This paper provides general information for carers and hospital staff on how to appropriately care for moderately or severely affected Myalgic Encephalomyelitis (M.E.) patients.

2. Hospital or carer notes for M.E. patients: M.E. patient's care form

As part of the 'Hospital and carer notes for M.E.' paper, a form is provided to be filled out by M.E. patients to give carers and hospital staff more information about the patient's illness severity and their specific disabilities and symptoms. If writing in the book directly is not advisable, carers or patients can also either download a copy of this form (free) from the HFME website, or fill in a photocopy of the appropriate pages from this book.

3. Why patients with severe M.E. are housebound and bedbound

Knowledge of some of the basics of how M.E. affects the body and the limitations of each patient are vital if you provide care for someone with M.E. or even if you make comments or have any type of input into the way the disease is managed, in order to avoid unnecessary suffering and disability. This paper provides a brief overview of this topic for carers, doctors or hospital staff, as well as friends and family members of patients.

4. Assisting the M.E. patient in managing relapses and adrenaline surges

5. Assisting the M.E. patient in managing relapses and adrenaline surges: Summary

6. Assisting the M.E. patient in having blood taken for testing

7. Assisting the M.E. patient in managing bathing and hair-care tasks

8. Assisting the M.E. patient in managing toileting tasks

9. The HFME 3 part ability scale for M.E. patients

The scales on this page are designed to be used by M.E. sufferers and their carers to measure improvements and changes over different aspects of M.E. over time. Because physical and cognitive ability and symptom severity are often not equally affected in every patient, this scale is divided into three parts.

10. The HFME ability scale for M.E. patients: Summary

11. The HFME ability and severity checklist

This paper provides a quick way for patients or carers to monitor the illness severity of M.E. and the severity of certain symptoms over time.

Hospital or carer notes for M.E. patients

 Patients with Myalgic Encephalomyelitis (M.E.) have a variety of specific care needs, some of which are well-known and common to a variety of other illnesses and others which are unique to M.E. and with which hospital staff or carers may be wholly unfamiliar.

Inappropriate care (even if well intentioned) can have serious consequences for M.E. patients in the short term and the long term, or even permanently. Knowledge of some of the basics about how M.E. affects the body is vital if you are in the position of providing care for someone with M.E. in order to avoid additional unnecessary suffering and disability. This paper provides a brief overview of this topic for hospital staff, carers or family members.

What is Myalgic Encephalomyelitis? How does it affect the body?

M.E. is a debilitating neurological (CNS) disease which has been recognised by the World Health Organisation since 1969 as a distinct organic neurological disorder. It can occur in both epidemic and sporadic forms and over 60 outbreaks of M.E. have been recorded worldwide since 1934.

M.E. is an acute onset neurological disease initiated by a virus (an enterovirus) with multi system involvement which is characterised by post encephalitic damage to the brain stem (hence the name 'Myalgic Encephalomyelitis'). M.E. is similar in a number of significant ways to diseases such as Multiple Sclerosis (M.S.), Lupus and Polio. M.E. can be extremely disabling; at least 30% of M.E. sufferers are severely affected and are almost completely (or completely) housebound and/or bedbound. Children as young as five can get M.E., as well as adults of all ages. M.E. has a similar strike-rate to M.S. and is a (potentially fatal) chronic/lifelong illness.

M.E. is primarily neurological, but because the brain controls all vital bodily functions, virtually every bodily system can be affected by M.E. Although M.E. is primarily neurological it is also known that the vascular and cardiac dysfunctions seen in M.E. are the cause of many of the symptoms and much of the disability associated with M.E., and that the well-documented mitochondrial abnormalities present in M.E. significantly contribute to both of these pathologies. There is also multi-system involvement of cardiac and skeletal muscle, liver, lymphoid and endocrine organs in M.E.

Thus M.E. symptoms are manifested by virtually all bodily systems including: cognitive, cardiac, cardiovascular, immunological, endocrinological, respiratory, hormonal, gastrointestinal and musculo-skeletal dysfunctions and damage. Myalgic Encephalomyelitis affects the brain, the heart, almost every bodily system and every cell of the body. One of the defining features of M.E. is an inability to maintain homeostasis.

All of this is not simply theory, but is based upon an enormous body of mutually supportive clinical information. These are well-documented, scientifically sound explanations for why patients are housebound or bedridden, profoundly intellectually impaired, unable to maintain an upright posture and so on (Chabursky et al. 1992 p. 20) (Hyde 2007, [Online]) (Hyde 2006, [Online]) (Hyde 2003, [Online]) (Dowsett 2001a, [Online]) (Dowsett 2000, [Online]) (Dowsett 1999a, 1999b, [Online]) (Hyde 1992 pp. x-xxi) (Hyde & Jain 1992 pp. 38 - 43) (Hyde et al. 1992, pp. 25-37) (Dowsett et al. 1990, pp. 285-291) (Ramsay 1986, [Online]) (Dowsett & Ramsay n.d., pp. 81-84) (Richardson n.d., pp. 85-92).

What all of this means in practice is that patients with M.E. have to be very careful with, or limit:

- Physical activity
- Cognitive activity

- Sensory input (exposure to light, noise, movement and vibration), and
- Orthostatic stress (maintaining an upright posture)

The main characteristics of the pattern of symptom exacerbations, relapses and disease progression (and so on) in M.E. include:

A. People with M.E. are unable to maintain their pre-illness activity levels. This is an acute (sudden) change. M.E. patients can only achieve 50%, or less, of their pre-illness activity levels post-M.E.
B. People with M.E. are limited in how physically active they can be but they are also limited in similar ways with; cognitive exertion, sensory input and orthostatic stress.
C. When a person with M.E. is active beyond their individual (physical, cognitive, sensory or orthostatic) limits, this causes a worsening of various neurological, cognitive, cardiac, cardiovascular, immunological, endocrinological, respiratory, hormonal, muscular, gastrointestinal and other symptoms.
D. The level of physical activity, cognitive exertion, sensory input or orthostatic stress needed to cause a significant or severe worsening of symptoms varies from patient to patient, but is often trivial compared to a patient's pre-illness tolerances and abilities.
E. The severity of M.E. waxes and wanes throughout the hour/day/week and month.
F. The worsening of the illness caused by overexertion often does not peak until 24 - 48 hours (or more) later.
G. The effects of overexertion can accumulate over longer periods of time and lead to disease progression, or death.
H. The activity limits of M.E. are not short term: a gradual (or sudden) increase in activity levels beyond a patient's individual limits causes relapse, disease progression or death in patients with M.E.
I. The symptoms of M.E. do not resolve with rest. The symptoms and disability of M.E. are not just caused by overexertion; there is also a base level of illness which can be quite severe even at rest.
J. Repeated overexertion can harm chances for future improvement in M.E. Patients who are able to avoid overexertion have repeatedly been shown to have the most positive long-term prognosis.
K. Not every M.E. sufferer has 'safe' activity limits within which they will not exacerbate their illness; this is not the case for the very severely affected.

In short, if patients with M.E. exceed their individual post-illness physical, cognitive, orthostatic and other limits, they will experience some combination of the following:

- A mild-severe (acute or delayed) worsening of one or more symptoms for hours, days or longer afterward
- A mild-severe (acute or delayed) worsening of virtually every symptom for hours, days or longer afterward
- A severe (acute or delayed) worsening of the base level of illness/disability for hours/ weeks/ months or even years afterward, or
- A permanent worsening of the base level of illness/disability (i.e. permanent physical damage is caused and chances for significant recovery are adversely affected or taken entirely)

It is also important to be aware that repeated or severe overexertion can result in the death of the M.E. patient. (Death in M.E. is most often caused by heart failure or multiple organ failure.)

What are the top 10 most obvious things you need to be aware of in providing care to an M.E. patient?

1. Reduce exposure to light
2. Reduce exposure to noise
3. Reduce/eliminate all non-essential visitors
4. Do not encourage patients to be more physically active (or upright longer) than they can easily tolerate
5. Try to schedule demanding tasks for the patient's best time of day as much as is possible
6. Try to reduce the patient's levels of cognitive exertion and sensory input

7. Be aware of any special dietary requirements
8. Be aware of the likelihood of negative drug reactions
9. Be aware of problems with sleep and the need for extensive rest
10. Be aware that these aforementioned relapses can be delayed, and that they can be very serious and prolonged

1. Reduce exposure to light

-Some patients will require the room to be completely dark (or very close to it), some will be fine so long as blinds and doors are kept closed, while other patients will fit somewhere in-between these two extremes.

2. Reduce exposure to noise

-At a minimum, doors and windows must be kept closed to reduce noise. Anyone entering the room must also take care to reduce or eliminate noise as much as possible, particularly if a patient has severe noise sensitivity.

-Open wards such as in emergency rooms are a DISASTER for M.E. patients. They will without exception cause months or more of severe relapse in the severely affected and may also cause a more immediate worsening of the overall condition and should be avoided if at all possible. Moderately affected patients may also relapse severely in an open ward. Sharing a room with another patient is also inappropriate for the severely affected M.E. patient and will cause a high level of increased pain and suffering and long term relapse.

-The problem here is not merely pain in the ears and painful or burning eyes. Even low levels of noise or light (and other sensory input) can cause a significant and prolonged worsening of the severity of the condition overall, as well as symptoms including seizures, severe mental confusion and inability to process even very simple information, episodes of paralysis, problems with proprioception, balance and so on. Pain levels can quickly soar to a 10/10 level even with moderate or brief noise or light exposure, and recovery can be prolonged.

3. Reduce/eliminate all non-essential visitors

-As well as reacting badly to the extra noise and light exposure caused by visitors, patients can also be made sicker by watching the movement of someone in the room, and by the extra demands made on the brain when talking and listening to speech is required.

-In the case of cleaners, these should be cancelled for the duration of the hospital stay, both for the reasons outlined above, and because many M.E. and M.S. patients have sensitivities to common chemicals used in cleaning products. (Exposure to these chemicals may merely trigger headaches but in some cases they can cause severe relapse.)

-It is counter-productive and ill-advised to do hourly 'obs' (pulse and blood pressure checks etc.) on a patient with severe M.E. as the repeated interruptions may cause them to deteriorate in both the short and the long term (or even permanently).

4. Do not encourage patients to be more physically active (or upright longer) than their bodies and hearts can easily tolerate

-Even sitting up in bed propped up by a few pillows counts as 'being upright' when someone is severely affected, and 30 seconds or a few minutes of being fully upright may be long enough to cause problems.

-Physical activity doesn't just include strenuous activity, but any movement. Even simple movements or stretching of the muscles can cause a worsening of the condition in the severely affected. Physical tasks may need to be broken up into many smaller tasks with long rest periods in-between.

-Some patients will require wheelchairs, but those who also have severe orthostatic problems (problems with being upright, including sitting) must not be put in wheelchairs at all and will need to be moved lying flat in bed (or lying flat on the back seat of a car) at all times. Even then travelling can still cause severe relapse.

5. Try to schedule demanding tasks for the patients best time of day as much as is possible

-Find out when the patient's best time of day is, and try to fit tasks in to that window as much as possible.

-Don't expect that a patient will necessarily be able to do the same things at different times of the day. Some tasks may only be possible at certain times of day, or after a long period of rest. Making a patient do difficult tasks at the time of day when they are at their most ill, can not only make the task much harder or impossible, but also cause a far worse relapse than if attempted at their 'most well' time of day.

-It is also important that you pay attention to visual and other cues that let you know how ill and how disabled an M.E. patient is at a particular time, and when is and isn't an appropriate time to engage them in conversation or to attempt potentially difficult tasks. For example, if the patient does not speak to you on either their walk to the bathroom or the walk from the bathroom back to their bed it is most likely that they are not well enough to speak at this time and so should *not* be engaged in conversation.

It should not be assumed that just because a patient can do one task or is out of bed, that they are feeling well or are at their best and are also capable of completing other tasks at the same time. Often the opposite is true: the fact that the patient is already doing something difficult leaves them *less likely* than usual to be feeling well and able, rather than more. As a general rule, if someone has severe M.E., please don't speak to them more than you absolutely have to, unless you are given clear encouragement by the patient at this time to do so. Even then, watch the patient carefully for signs that they are starting to become ill from this exertion and end the conversation as soon as possible when this occurs.

-If possible, it would be helpful for carers to be on the lookout for signs that the patient is pushing themselves to the point of relapse and to make the patient aware of this and of the need for them to rest immediately. These signs may be noticed by an observer more quickly and more easily than by the patient.

Signs may include: talking excessively and very fast (due to bursts of adrenaline which can be released when the body is in severe physiological difficulty and unable to cope), talking in a very stilted way with large pauses between words, slow and slurred speech, the feet tuning purple or blue when the patient has been upright too long, the face turning white or the facial expression becoming blank (the person may also be 'slack-jawed' and have their mouth open), excessive blinking or an inability to keep the eyes open at all, excessive water drinking (which can indicate that the patient's body is trying to reduce cardiac insufficiency by increasing blood volume), sighs or grimaces of pain, increased clumsiness and mental confusion, and so on. Signs will vary depending on the person.

6. Try to reduce the patient's levels of cognitive exertion and sensory input

-Sensory input includes; light, noise, movement, touch and also vibration.

-Travelling by car can be excruciating with severe M.E. and can cause a severe and prolonged worsening of neurological, cardiac and other problems. Even being lifted from one bed to another can cause relapse.

-Cognitive exertion includes talking and listening to speech, reading and writing, watching TV, listening to music and so on. Talking as well as listening to speech can be very difficult or impossible. Even speaking to someone with M.E. for longer than they can easily cope with (even if you aren't forcing them to reply) can be disastrous and cause relapse or a deterioration of their overall condition. Cognitive tasks may need to be simplified and broken up into many smaller tasks with long rest periods in-between.

-Some severely affected patients are unable to maintain consciousness for more than short periods at a time. Some may only be properly conscious for a few hours a day or less. Sometimes consciousness cannot be maintained for more than 10 minutes or so consecutively (or less). Trying to force these patients into consciousness for longer periods can only be counter-productive, unfortunately. It can quickly make the problem even worse. (Aside from certain nutrients and other treatments, what will help improve this condition most is rest.)

-Even mildly cold or warm weather (or room temperatures) can cause severe problems (and suffering) for many M.E. patients; particularly warm or hot weather. If patients become cold it can exacerbate joint stiffness and pain, and beyond a certain point the body is unable to make itself warm again without something like a warm bath as with M.E. the brain and body have lost the ability to properly regulate temperature (to maintain homeostasis). Warm or hot weather is tolerated poorly by M.E. patients and can easily cause severe and prolonged relapse of all symptoms and an increased loss of ability (including cognitive abilities). Because M.E. makes the body unable to react and adapt as it should to warm and cold temperatures, patients must manually be able to make sure they do not get too hot nor too cold. Air conditioning is vital in summer, and blankets, a heater and possibly also warm baths or an electric heat pack are needed in winter. Some patients also have such severe problems with temperature control that they are unbearably cold even in the middle of summer, or they go from very hot to very cold over and over again from one 5 minute period to the next and so need to continually adjust blankets etc. to try to cope.)

7. Be aware of any special dietary requirements

-Patients will often be intolerant of a large variety of foods. Some may also have food allergies.

-There may also be strict requirements – due to the metabolic problems seen in M.E. – that a patient eats every 2 or 3 hours (or even more often) and that meals or snacks are high in protein or fat and low in sugar and carbohydrate to prevent relapse. High sugar or high carbohydrate foods are often very poorly tolerated by M.E. patients.

-Some patients will require assistance from a carer to eat (or tube feeding in severe cases). Problems with swallowing can also make eating or drinking difficult or impossible for the M.E. patient.

8. Be aware of the likelihood of negative drug reactions

-M.E. patients can react badly to almost every type of drug; particularly those which act upon the CNS. Some severely affected patients are unable to tolerate any drugs or over the counter vitamins and other supplements at all, although many will have found a small number of products, through much trial and error, that they can tolerate.

-Negative effects from taking certain medications can range from headaches and feelings of being poisoned, to a severe worsening of the overall condition which may be prolonged.

-All new medications should be started one at a time and at very low doses (e.g. 1/10th of a standard dose)

-Patients may also react badly to the chemicals contained in many personal care products. If this sensitivity is very severe, visitors must avoid wearing these products as much as possible before visiting. Chemicals often used in building or cleaning can cause pain, headaches and other symptoms in some patients, as can exposure to mouldy environments. An M.E. sufferer may be adversely affected by a level of chemicals or mould which is not detectable, or only barely detectable, by a healthy person. Not every M.E. patient is affected significantly by chemical and mould exposures but for some this is a severe problem.

9. Be aware of the need for extensive rest and problems with sleep

-Patients with M.E. need a lot of rest, but often find it impossible to get much sleep or find initiating sleep very difficult, or can only achieve a very low quality of sleep or sleep only for short periods at a time.

-It may take some patients 4 or more hours to initiate sleep. Being interrupted with noise or light or visitors during this time may make that period even longer, or prevent the initiation of sleep altogether. Even low level noise can sometimes wake M.E. patients who cannot achieve normal deep sleep.

-Some patients cannot ever sleep for more than a few hours at a time post-M.E., so they need to be left alone as much as possible in order that they get these much needed sleep periods. Sleep doesn't necessarily help M.E. symptoms much – often patients feel just as ill or even much worse on waking than they did before they went to sleep – but missed sleep causes severe worsening of symptoms/disability. The way it feels to have M.E. *and* sleep deprivation is horrific, particularly when M.E. is severe.

-Rest periods are very important in M.E. (including short micro-rests). Patients may be made far more ill by tasks which do not allow suitable rest periods. Even talking very fast with no pauses to someone with severe M.E. can cause them to become more ill and to not be able to understand what is being said, for example.

10. Be aware that these relapses can be significantly delayed (so they are not always visible on superficial examination), and that they can be very serious and prolonged - or even fatal in a minority of cases

-Don't make superficial judgements of a patient's ability levels. If you want to know how a patient is feeling or if they can or can't do a certain task, just ASK THEM!

- People with M.E. are very highly motivated to be as active as they possibly can be (as anyone would be with so many restriction on their lives), but they know that if they push themselves to do more than their bodies can handle, the end result will be a huge LOSS of ability levels, and a higher level of suffering, and so this is not in their best interests. People with M.E. must be proactive in one respect: by carefully staying within their post-M.E. limits. This also gives the patient the best chance for their best possible outcome.

-Do take the risk of relapse, and the patient's unwillingness to become far more ill for days, weeks or longer, very seriously. Many M.E. patients are suffering in a fairly extreme way already, and their lives are so painful and limited as to almost be unbearable, without any additional worsening of the condition.

Conclusion

Just do the best you can. Achieving all of these tasks perfectly all the time may not be possible. It's a lot to take in and a lot to think about all at once, but everything that you can do to reduce the relapse from a hospital stay will make a real difference and be much appreciated. There is a huge difference between a 2 month long relapse and a 6 month relapse; between symptoms worsening during this time to a 7/10 level rather than a 10/10 level; between a short-term and a permanent worsening of symptoms.

(M.E. patients appreciate what a hassle it is to accommodate the demands of M.E. only too well. M.E. is an acute onset disease. Those of us who have M.E. went from being normal and healthy one day to having to cope with great limits and disabilities the next, even from one hour to the next. M.E. patients understand that M.E. is very unforgiving, overwhelming and a huge hassle to deal with on just about every level; we understand the issues carers grapple with.)

Following this text are some additional forms about specific symptoms and disabilities that patients may or not want (or be well enough) to fill out in order to give you more information about their needs, where this is appropriate. Thank you for taking the time to read this paper.

Additional notes on this text:

Note that hospital trips (or any travelling out of the house) should be an absolute last resort for patients with severe M.E. because of the enormous price they pay for such trips. It should be avoided wherever possible. It's is counter-productive and cruel. This extreme level of suffering is not short term either. It is very common for severely affected patients to spend 6 months, 12 months, several years or longer recovering from a hospital trip. Some never do recover. Again, there have also been cases where an M.E. patient has left hospital only to go home and die.

People with severe M.E. are some of the most vulnerable members of society and they deserve and desperately need appropriate care; care given in the home as much as possible. It is unreasonable that these already very severely ill patients have to be made more severely ill to get the basic care they need, most of which could easily be administered at home at an immensely reduced physical cost to the patient. For more information see: Why patients with severe M.E. are housebound and bedbound.

Acknowledgments
Thanks to Lesley Ben and Emma Searle for editing this paper.

M.E. patient care form

A form to be filled out by the M.E. patient for the benefit of hospital staff and carers

Name: _____   Date: _____

(*Patient note:* Just fill in as much of the form as possible or as much as you feel is necessary. It doesn't matter if you can't read or fill-in all of it. If you are very ill, just circle those statements which are most important and leave the rest, or perhaps give this form to a knowledgeable family member to fill out on your behalf or as per your verbal instructions. Parents of a child with Myalgic Encephalomyelitis (M.E.) may need to assist the child in filling this form out. Where multiple options are given, either circle the most appropriate answer or cross out any which do not apply.)

1. Photophobia (sensitivity to light exposure)

My problems with photophobia are mild/moderate/severe/very severe
To minimise the pain and disability associated with this symptom I need:

_____

_____

_____

_____

2. Noise sensitivity

My problems with noise are mild/moderate/severe/very severe
The noises which bother me most are:

_____

_____

_____

The time of day my noise sensitivity is at its worst is morning /noon /afternoon /evening/all day
Voices can also be very painful for me. Could you please: whisper at all times and/or don't speak to me until afternoon /late afternoon/ evening or until I tell you that I am well enough to speak.
To cope with this problem I need:

_____

_____

3. Visitors

Having visitors causes me mild/moderate/severe/very severe relapse

Visitors to my room are fine, but could you just please carefully limit noise and/or light
Please limit all unnecessarily visitors to my room: yes/no
Please stop all unnecessarily visitors to my room: yes/no
Notes:

_____

_____

_____

_____

4. Physical activity and orthostatic stress

I am limited with regards to physical activity in the following ways:
- ❑ I need care 24 hours a day.
- ❑ I need assistance for almost everything. Unaided I cannot:

_____

_____

_____

_____

- ❑ I need minor/moderate/total assistance with dressing, eating, food preparation, toileting, bathing, personal care and hygiene, standing. I need:

_____

_____

_____

_____

I am limited with how much I can be upright, in the following ways:
- ❑ I can be upright for around ___ minutes/hours a day, but for no more than ___ minutes / hours at a time (only at times of my choosing, as I must have time to rest from last period standing before I can do it again).
- ❑ I cannot sit or be standing up at all.
- ❑ I have to eat, drink, toilet, bathe, dress lying down.
- ❑ My orthostatic problems are mild /moderate /severe. If I am upright for longer than I should be, what happens is the above , so I need you to:

_____

_____

_____

_____

5. Changes to my illness severity/abilities over the day

My best time/s of the day is/are usually:

_____

_____

My worst time/s of the day is usually:

___

___

6. Problems with cognitive tasks and sensory input

Some people with M.E. have problems with speaking. This is a mild /moderate /severe problem for me:
- ❑ All the time.
- ❑ At certain times of the day, or after I have been active in some way.
- ❑ Sometimes.
- ❑ Occasionally

The problem I have is (tick all boxes which apply):
- ❑ I can't speak at all.
- ❑ I can speak only a few words at a time.
- ❑ I can speak only at great physical cost (i.e. pain and/or relapse).
- ❑ I can only speak very quietly.
- ❑ I slur my words and am hard to understand.
- ❑ Sometimes I cannot speak at all and the only way I can communicate is through using hand signals

Could you please help me with this by:

___

___

___

___

The cognitive problems I have are mild/moderate /severe/very severe
- ❑ I need assistance with everything
- ❑ I often need help understanding things
- ❑ I occasionally need help understanding things if very unwell
- ❑ I often need help remembering things.
- ❑ I cope better with written information than audio
- ❑ I cope better with audio information than text
- ❑ I need things repeated to me many times over
- ❑ I need to record conversations/ take notes so that I remember them
- ❑ I forget things sometimes /often /very often.
- ❑ I need things explained to me in a simple way sometimes/often/always
- ❑ I find it very hard/ impossible to read or to understand speech at times.
- ❑ I often have difficulty maintaining consciousness for more than _____ minutes/hours at a time.
- ❑ I often have difficulty paying attention and thinking for more than _____ minutes/hours at a time or for more than _____ minutes/hours a day.
- ❑ I can usually watch ___ minutes/hours of TV a day, for ___ minutes/hours at a time, on a good day.
- ❑ I can listen to the radio or music for ___ minutes/hours at a time, for ___ minutes/hours a day, on a good day
- ❑ I can usually read for _____ minutes/hours a day, on a good day
- ❑ I can usually talk for _____ minutes/hours a day, on a good day

I would appreciate it if you could help me with my cognitive problems by:

___

___

___

___

I'm sensitive to temperature in the following ways:
- ❏ Very warm temperatures make me ill.
- ❏ Even mildly warm temperatures make me ill.
- ❏ If I get even slightly cold it is very unpleasant and I soon become extremely cold for a long time afterwards (or pass out in a 'cold fever').
- ❏ I only do well in a very narrow range of temperatures, slightly cold or warm weather makes me ill.
- ❏ Cold weather increases the problems with my joints and my pain levels.

I am sensitive to touch in a mild/moderate/severe way. Could you help me with this by:

....

....

....

....

I am sensitive to vibration in a mild/moderate/severe way. Could you help me with this by:

....

....

....

....

7. Special dietary requirements

My diet is affected by my illness in the following ways:
- ❏ I require small meals every __ minutes/hours (due to metabolic problems)
- ❏ Foods which I am intolerant of in some way include: stimulants, sweeteners (sugar, dextrose, glucose, fructose, splenda, aspartame and saccharin), additives (artificial colours, flavours, preservatives, MSG), foods from the nightshade family (potato, capsicum, eggplant and tomato), dairy products, fruit (may be difficult to digest and the high levels of fructose can trigger hypoglycaemia and other problems), raw foods (may be difficult to digest), fermented and mouldy foods and foods containing yeast or wheat, acidic foods, nuts and soy.
- ❏ I also need to avoid the following foods:

....

....

....

....

- ❏ I need my food cut up for me sometimes/often/always
- ❏ I require assistance to eat some/most /all meals.
- ❏ I require a low glycaemic index diet (foods which release carbohydrate slowly).
- ❏ I require a low glycaemic load diet (foods which are low in carbohydrate).
- ❏ I require liquid meals/tube feeding.
- ❏ I need special care to be taken that you wash your hands thoroughly before you prepare my food, and I need all fruit and vegetables etc. to be well washed to remove pesticide residues as much as possible.
- ❏ Due to difficulties eating / swallowing, I need to avoid foods like:

....

....

....

....

Other notes on diet:

_____

_____

_____

_____

_____

8. Negative drug and chemical reactions

I have sensitivities to certain drugs (and other medications) that are usually mild/moderate/severe.

❑ I can take most medications, but have a problem with:

_____

_____

_____

❑ I have problems with many different medications.
❑ I have a problem with almost every medication; so far I can't take anything at all.
❑ I can take a variety of over-the-counter vitamins and herbs, but tend to have more problems with prescription drugs (particularly those which act on the CNS).
❑ Please read the paper Anaesthesia and M.E. before I have my operation so that you are aware of some of the things to look out for with regard to anaesthesia and M.E.

Other notes:

_____

_____

_____

I have sensitivities to chemicals and airborne allergens that are: mild /moderate/severe. I am sensitive to/precautions that I have to take because of this problem include:

_____

_____

_____

If I don't take these precautions, what happens is:

_____

_____

_____

9. Rest and problems with sleep

My sleep is affected by M.E. in the following ways:
❑ Insomnia / broken sleep.
❑ Reversed sleep/wake cycle (I usually go to sleep at _____ and wake at _____)
❑ I have sleep paralysis: occasionally/ often /always, it usually lasts for _____
❑ I sleep very lightly and need things to be as quiet as possible while I sleep.

Notes:

_____
_____
_____
_____

10. Miscellaneous

A. Emotional stress: Excessive emotional stress can cause my condition to worsen, as it requires the body to work harder physically to cope. This is usually a mild/moderate/severe problem for me. Could you please try to avoid:

_____
_____
_____
_____

B. Secondary infections: Some people with M.E. are very susceptible to secondary infections i.e. colds/flu etc. These infections can also last much longer or be more severe than usual.
❏ This is a mild/moderate /severe problem for me. It would be great if you could please:

_____
_____
_____

C. What this trip to hospital will mean for me: Judging from my many past experiences with overexertion, when I get home from hospital I will likely be mildly/ severely/very severely more ill than usual. This will probably last for

_____
_____

❏ The reason I am so severely affected in the first place is because I was encouraged or forced to be more active than my body could cope with in the early stages of my illness. When I was first ill I was only moderately affected.
❏ The reason I am so severely affected is because of inappropriate medical care. When I was first ill I was only moderately affected.

D. Other symptoms not mentioned: Other symptoms I have that it is important that you know about include:

_____
_____
_____
_____

E. My worst symptoms: My top 5 problems, disabilities or symptoms would be:

_____
_____
_____

In short, please take care most of all to:

_____

_____

_____

Thank you.

Why severe M.E. patients are housebound and bedbound

COPYRIGHT © JODI BASSETT NOVEMBER 2008. UPDATED JUNE 2011. FROM WWW.HFME.ORG

 Knowledge of some of the basics of how Myalgic Encephalomyelitis (M.E.) affects the body and the limitations of each patient are vital if you provide care for someone with M.E. This knowledge is also necessary even if you make comments or have any type of input into the way the disease is managed, in order to avoid additional unnecessary suffering and disability. This is so important with M.E. because inappropriate care, comments and advice or pressure for M.E. patients to do certain things (even if well intentioned) can have serious consequences for the patients in the short term and the long term, or even permanently.

This paper provides a brief overview of this topic for friends and family members, and also for carers, doctors or hospital staff.

Why are some severely affected M.E. patients housebound?

This is a question that severe M.E. patients are sometimes asked. The short answer to this question is:

A. They are simply too ill and disabled to leave the house. This task is physically impossible for them due to the severity of their illness, or:
B. They are physically able to leave the house, but it would be unwise for them to do so. In the short term this type of overexertion causes even more severe suffering than is already experienced daily (and may already be at an unbearable level). Even worse, this extreme additional loss of quality of life and ability can and does persist for a long time afterward.

It is very common for severely affected patients to spend 2 months, 6 months, 12 months or even several years or longer recovering from a hospital trip (etc.). For example, some patients still have not regained their previous low-level of health 2 or 4 years after a trip to hospital. Some never do recover, and for some patients the overexertion is so severe as to be fatal.

Severe overexertion also ruins a patient's chances for significant (or any) future recovery, and can cause permanent physical damage.

Severely affected M.E. patients may also sometimes be asked questions such as:

- 'Why are you bedbound, or wheelchair-bound?
- 'Why are you almost completely housebound or bedbound?'
- 'Why have you had to stop studying or working?'
- 'Why can't you do all the tasks of daily living for yourself?'
- 'Why can't you use the phone, or watch TV?'

The answer to each of these questions is the same, it's just a difference of degree. Some tasks are physically impossible for some sufferers, and others are possible but unwise. Sometimes tasks can be done in a controlled way, and limited as to frequency and/or duration. In other words, the activities need to be carefully 'rationed.'

That is really all there is to it. A person with M.E. doesn't do certain things they would like to, because they are either too ill to do them, or because it would reduce their ability and quality of life for months or years afterward. They may lose any chance at significant recovery by pushing themselves to do something that their severely damaged bodies can't cope with.

That is the short answer. If you'd like more detail on all of these points, and some more M.E.-specific medical information and treatment and management guidelines, then please read on.

What is M.E.? How does it affect the body?

Myalgic Encephalomyelitis is a debilitating neurological (CNS) disease which has been recognised by the World Health Organisation since 1969 as a distinct organic neurological disorder. It can occur in both epidemic and sporadic forms and over 60 outbreaks of M.E. have been recorded worldwide since 1934.

M.E. is an acute onset neurological disease initiated by a virus (an enterovirus) with multi system involvement which is characterised by post encephalitic damage to the brain stem (hence the name 'Myalgic Encephalomyelitis'). M.E. is similar in a number of significant ways to diseases such as Multiple Sclerosis (M.S.), Lupus and Polio. M.E. can be extremely disabling; at least 30% of M.E. sufferers are severely affected and are almost completely (or completely) housebound and/or bedbound. Children as young as five can get M.E., as well as adults of all ages. M.E. has a similar strike-rate to M.S. and is a (potentially fatal) chronic/lifelong illness.

M.E. is primarily neurological, but because the brain controls all vital bodily functions virtually every bodily system can be affected by M.E. Although M.E. is primarily neurological it is also known that the vascular and cardiac dysfunctions seen in M.E. are also the cause of many of the symptoms and much of the disability associated with M.E., and that the well-documented mitochondrial abnormalities present in M.E. significantly contribute to both of these pathologies. There is also multi-system involvement of cardiac and skeletal muscle, liver, lymphoid and endocrine organs in M.E.

M.E. symptoms are manifested by virtually all bodily systems including: cognitive, cardiac, cardiovascular, immunological, endocrinological, respiratory, hormonal, gastrointestinal and musculo-skeletal dysfunctions and damage. M.E. affects the brain, the heart, almost every bodily system and every cell of the body. One of the defining features of M.E. is an inability to maintain homeostasis.

All of this is not simply theory, but is based upon an enormous body of mutually supportive clinical information. These are well-documented, scientifically sound explanations for why patients are housebound or bedridden, profoundly intellectually impaired, unable to maintain an upright posture and so on (Chabursky et al. 1992 p. 20) (Hyde 2007, [Online]) (Hyde 2006, [Online]) (Hyde 2003, [Online]) (Dowsett 2001a, [Online]) (Dowsett 2000, [Online]) (Dowsett 1999a, 1999b, [Online]) (Hyde 1992 pp. x-xxi) (Hyde & Jain 1992 pp. 38 - 43) (Hyde et al. 1992, pp. 25-37) (Dowsett et al. 1990, pp. 285-291) (Ramsay 1986, [Online]) (Dowsett & Ramsay n.d., pp. 81-84) (Richardson n.d., pp. 85-92).

What all of this means in practice is that patients with M.E. have to be very careful with, or limit:

* Physical activity
* Cognitive activity
* Sensory input (exposure to light, noise, movement and vibration), and
* Orthostatic stress (maintaining an upright posture)

The main characteristics of the pattern of symptom exacerbations, relapses and disease progression (and so on) in M.E. include:

A. People with M.E. are unable to maintain their pre-illness activity levels. This is an acute (sudden) change. M.E. patients can only achieve 50%, or less, of their pre-illness activity levels post-M.E.

B. People with M.E. are limited in how physically active they can be but they are also limited in similar way with; cognitive exertion, sensory input and orthostatic stress.

C. When a person with M.E. is active beyond their individual (physical, cognitive, sensory or orthostatic) limits this causes a worsening of various neurological, cognitive, cardiac, cardiovascular, immunological, endocrinological, respiratory, hormonal, muscular, gastrointestinal and other symptoms.

D. The level of physical activity, cognitive exertion, sensory input or orthostatic stress needed to cause a significant or severe worsening of symptoms varies from patient to patient, but is often trivial compared to a patient's pre-illness tolerances and abilities.

E. The severity of M.E. waxes and wanes throughout the hour/day/week and month.

F. The worsening of the illness caused by overexertion often does not peak until 24 - 72 hours (or more) later.

G. The effects of overexertion can accumulate over longer periods of time and lead to disease progression, or death.

H. The activity limits of M.E. are not short term: a gradual (or sudden) increase in activity levels beyond a patient's individual limits can only cause relapse, disease progression or death in patients with M.E.

I. The symptoms of M.E. do not resolve with rest. The symptoms and disability of M.E. are not just caused by overexertion; there is also a base level of illness which can be quite severe even at rest.

J. Repeated overexertion can harm the patient's chances for future improvement in M.E. M.E. patients who are able to avoid overexertion have repeatedly been shown to have the most positive long-term prognosis.

K. Not every M.E. sufferer has 'safe' activity limits within which they will not exacerbate their illness; this is not the case for the very severely affected.

In short, if patients with M.E. exceed their individual physical, cognitive, orthostatic and other limits, they will experience some combination of the following:

- A mild-severe (acute or delayed) worsening of one or more symptoms for hours, days or longer afterward
- A mild-severe (acute or delayed) worsening of virtually every symptom for hours, days or longer afterward
- A severe (acute or delayed) worsening of the base level of illness/disability for hours/ weeks/ months or even years afterward, or
- A permanent worsening of the base level of illness/disability (i.e. permanent physical damage is caused and chances for significant recovery are adversely affected or lost entirely. Painstaking gains made slowly over many months or years may also be lost.)

It is also important to be aware that repeated or severe overexertion can result in the death of the M.E. patient. (Death in M.E. is most often caused by heart failure or multiple organ failure.) (Bassett, 2010, [Online])

For these reasons, it is vitally important that patients are allowed to judge *for themselves* how much activity it is safe and wise for them to attempt. Patients are the best judges of their own limits, and patients' judgements must not be over-ruled. Patients should never be advised, encouraged or forced to be more active than their severely damaged bodies can handle; these decisions cannot safely or ethically be made by any third party.

What are the problems for severe M.E. patients being out of their bed or home?

"How is the M.E. patient being overexerted and made more ill if they are transported somewhere while lying down?," or " how can just a few minutes or hours out of bed possibly make the patient more ill long-term?" a healthy person might ask.

It is common for people dealing with M.E. patients to pay close attention to the fact that a patient with M.E. has to limit physical overexertion, but to not fully understand that excessive sensory input and cognitive exertion and other factors can make the patient just as ill as excess physical activity. These factors are also much harder to minimise. For example:

- It is impossible to avoid additional cognitive stimulus during a trip out of the house. Whether it is looking at new environments, or having to listen to speech or being asked to answer questions and make decisions or just being asked to speak at all, all of these things can be unbearable for the severe M.E. patient and cause severe problems in the short and long term.

- Sensory input such as excessive (or even low level) noise, light and even vibration or a sense of movement (as felt when travelling by car or ambulance) can be unbearable and extremely painful for the severe M.E. patients and cause significant problems. The problem here is not merely pain in the ears and painful or burning eyes. Even low levels of noise or light (and other sensory input) can cause a significant and prolonged worsening of the severity of the condition overall, as well as symptoms including seizures, mental confusion and inability to process even very simple information, episodes of paralysis, problems with proprioception, balance and so on. Pain levels can quickly soar to a 10/10 level even with moderate or brief noise or light exposure, and recovery can be prolonged. Travelling by car is excruciating with severe M.E. and can cause a prolonged, *or permanent*, worsening of neurological, cardiac and other problems. It can also cause death (see section/question 1 below).

Note too that travelling by car causes relapse even if light, noise and vibration are minimised as much as possible. The problem isn't *just* excess sensory input. Even then, as one M.E. patient explains it, it is also the exertion of <u>movement through space</u> that leaves severe M.E. patients 'in a coma-like state' and feeling as if they're 'going into total organ failure' during and after travelling.

- A patient's <u>inability to be upright</u> for any amount of time can be very severe. Often trips out of the house, even where a patient is transferred by bed almost entirely, still require a patient to sit up for short periods which can be unbearable for the very ill and cause significant problems in the short and long term. Even sitting up in bed propped up by a few pillows counts as 'being upright' when someone is severely affected, and even a few minutes of being upright may be long enough to cause major problems.

- <u>Exposure to warm or cool temperatures</u> can also cause sometimes acute problems (as patients with M.E. have a loss of thermoregulation).

- <u>Exposure to chemicals</u> in new environments (from common personal care products worn by others, to chemicals used in building or cleaning) can cause pain, headaches and other symptoms in some patients, as can <u>exposure to mouldy environments</u>. An M.E. sufferer may be adversely affected by a level of chemicals or mould which is not detectable, or only barely detectable, by a healthy person. Not every M.E. patients is affected significantly by chemical and mould exposures but for some this is a significant problem.

- Patients with M.E. often also have very <u>restricted diets</u> (due to digestion problems, food allergies and intolerances etc.), and problems with going for even a few hours, or more than half an hour in some cases, without food (as with other patients with severe metabolic/mitochondrial disorders). There is also a need to have continual access to adequate <u>water</u>. Trips out of the house that don't accommodate these needs can make the patient very ill.

So as you can see, merely protecting the patient from physical overexertion is not enough by itself to make an activity safe for an M.E. patient. It is more complicated than that unfortunately.

One of the main misconceptions is that while walking a few steps requires additional bodily resources and cardiac output, time spent thinking, looking, listening or experiencing other sensory stimuli does not. This is not the case. Not only physical effort, but also cognitive effort, requires additional resources which an M.E. patient may not have. The brain contains some 100 billion neurons connected to 10,000 relay stations and this enormous electrical activity creates a massive need for energy and other bodily resources. The brain uses up to 25% of the entire body's demand for glucose, 25% of the blood pumped from the heart goes to the brain and the brain also needs 25% of the body's oxygen supply. (Blood supplies nutrients like glucose, protein, trace elements, and oxygen to the brain.) So of course, every extra second of 'electrical activity' – every thought, every feeling, every noise heard or sight seen – requires additional cardiac output, makes additional oxygen and glucose demands, and so on, in just the same way as does a physical activity such as walking; if not more so. So in addition to physical activity, the list of things that can cause similar severe relapse in M.E. patients also includes cognitive exertion, sensory input and orthostatic stress; anything that makes the body work harder or have to adjust in some way (Dowsett n.d. d, [Online]).

Again, that is why hospital trips (or any travelling out of the house) should be an absolute last resort for patients with severe M.E. and should be avoided wherever possible. It is counter-productive and cruel.

People with severe M.E. are some of the most vulnerable members of society and they deserve and desperately need appropriate care; care given in the home as much as possible.

It is unreasonable that these already very ill patients have to be made so much more ill to get the basic care they need, most of which could easily be administered at home at an immensely reduced physical cost to the patient.

Advice for carers

If there is a genuine need for a trip out of the house there are things that can and must be done to help minimise the harm caused. So what are the top 10 most obvious things that need to be considered by anyone providing care to an M.E. patient on a daily basis, whether at home, in transit or during a short trip to hospital?

1. Reduce exposure to light
2. Reduce exposure to noise
3. Reduce/eliminate all non-essential visitors
4. Do not encourage patients to be more physically active (or upright longer) than they can easily tolerate
5. Try to schedule demanding tasks for the patient's best time of day as much as is possible
6. Try to reduce the patient's levels of cognitive exertion and sensory input
7. Be aware of any special dietary requirements
8. Be aware of the likelihood of negative drug reactions
9. Be aware of problems with sleep and the need for extensive rest
10. Be aware that these aforementioned relapses can be delayed, and that they can be very serious and prolonged

Each of these points is expanded upon in the text: Hospital or carer notes for M.E.

It's a lot to take in all at once, but everything that you can do to reduce the relapse from a hospital stay – or even better, avoid a hospital stay completely – will make a real difference and be much appreciated. Just do your best.

There is a huge difference between a 2 month long relapse and a 6 month relapse; between symptoms worsening during this time to a 7/10 level rather than a 10/10 level; between a short-term and a permanent worsening of symptoms.

(M.E. patients appreciate what a hassle it is to accommodate the demands of M.E. only too well. M.E. is an acute onset disease. Those of us who have M.E. went from being normal and healthy one day to having to cope with great limits and disabilities the next, even from one hour to the next. M.E. patients understand that M.E. is very unforgiving, overwhelming and a huge hassle to deal with on just about every level; we understand the issues carers grapple with.)

Conclusion

Some tasks are physically impossible for some M.E. sufferers, and others are possible but unwise. Sometimes difficult tasks can be done so long as it is in a controlled way; and strictly limited as to frequency and/or duration. Another way to say this is that some activities need to be very carefully 'rationed.' In addition, some tasks are only possible at the patient's best time of day, or with a period of rest beforehand (lasting minutes, hours or days or longer) or can only be completed if the task is modified in some way, or with assistance from a carer.

Activities that would be trivial for healthy people – including being out of bed or leaving the house for brief periods – can have disastrous consequences for patients with severe M.E. Consequences can include extremely severe and prolonged relapses, additional disability and suffering, permanent bodily damage and death.

Again, it is vitally important that M.E. patients are allowed to judge *for themselves* how much activity it is safe and wise for them to attempt. Patients are the best judges of their own limits, and patients' judgements must not be over-ruled. Patients should never be advised, encouraged or forced to be more active than their severely damaged bodies can handle; these decisions cannot safely or ethically be made by any third party.

If a patient says they *cannot* or *should not* do something: then family, friends, doctors, carers and hospital staff *must listen*. Thank you for taking the time to read this paper.

Extra questions and answers section:

1. Can severe M.E. patients really die just from being forced out of bed, or to leave the house etc.?
2. How important is appropriate rest in M.E.?
3. What does 'rest' mean exactly in this context?
4. What does 10/10 pain and suffering mean in this context?
5. What is homeostasis?
6. Is M.E. a stable illness?
7. Why is M.E. not the same thing as 'CFS'?

1. Can severe M.E. patients really die just from being forced out of bed, or to leave the house etc.?

Of course I cannot show you a double blind controlled study where 25 severe M.E. patients were taken out of the house, and 25 were left at home to rest and show you how many of those moved from home died and how many didn't. This subject is a difficult one to research (even without the political problems obstructing genuine M.E. research) as it involves making patients very much more ill or killing them, which is obviously something no ethical and knowledgeable researcher would want any involvement with. However, we can look at the facts of M.E., research, the experience of moderately ill patients and M.E. fatalities and draw some conclusions.

Firstly, we know that M.E. can be fatal. Deaths from M.E. are well documented. For example, M.E. expert Dr Elizabeth Dowsett states: '20% have progressive and frequently undiagnosed degeneration of cardiac muscle which has led to sudden death following exercise' (Dowsett & Ramsay et al. 1990) (Dowsett 2000, [Online]) (Dowsett a, [Online]).

Deaths from severe CNS abnormalities are also described, as well as deaths caused by multiple organ failure or pancreatic failure. The term Myalgic Encephalomyelitis itself was created in UK in 1956 after doctors saw evidence of these abnormalities during autopsy on brains of patients who had died from M.E.

M.E. expert Dr Byron Hyde explains:

> I have some M.E. patients with a circulating red blood cell volume less than 50% of expected and a very large number with the range of 60% to 70%. What this test means is that blood is pooling somewhere in the body and that this blood is probably not available for the brain. When blood flow to the heart decreases sufficiently, the organism has an increased risk of death. Accordingly, the human body operates in part with pressoreceptors that protect and maintain heart blood supply. When blood flow decreases, pressoreceptors decrease blood flow to non-cardiac organs and shunt blood to the heart to maintain life. This, of course, robs those areas of the body that are not essential for maintaining life and means the brain, muscles, and peripheral circulation are placed in physiological difficulty (Hyde 2003, [Online]).

This physiological difficulty is exacerbated by physical and mental activity and orthostatic stress.

Dr Paul Cheney explains that when M.E. patients stand up, they are on the edge of organ failure as their cardiac output has dropped to the extremely low level of 3.7 litres per minute, a 50% drop from the normal output of 7 litres per minute. Without exception, says Cheney, every M.E. patient 'is in heart failure.'

Cardiac and vascular abnormalities have been documented from the earliest outbreaks of M.E. to the present day. Recent research shows that mitochondrial and other dysfunction leads to diastolic dysfunction and reduced stroke volume/low cardiac output in M.E. – and that certain levels of orthostatic stress and physical and mental activity exacerbate this cardiac insufficiency. Dr Cheney explained recently that because it takes more metabolic energy for the heart to relax and fill with blood than it does for it to squeeze and pump blood, the hearts of people with M.E. don't fill with the proper amount of blood before they pump which is what causes the reduced cardiac output and many of the symptoms of M.E. (and much of the disability of M.E.). So the tachycardia – fast heart rate – often seen in M.E. in response to orthostatic stress is actually compensating for low stroke volume to help increase cardiac output. The heart doesn't fill with enough blood before each beat of the heart so it is forced to beat faster to try to make up some of the shortfall, but people with M.E. are still left with reduced cardiac output which leaves them very ill and disabled. If this problem is critical enough it can result in death (Cheney 2006, [video recording]).

As one advocate explains: 'Cardiac output is sometimes too low to meet the demands of movement, and any attempt to exert oneself beyond one's own capacity for cardiac output - that is when demand exceeds cardiac capacity - would indeed result in death. Studies on dogs have shown that when the demands of the body exceed cardiac output by even 1%, the organism dies. M.E. patients [must] reduce demand and reduce their exertion level to stay within the bounds of their low cardiac output to stay alive' (MESA, 2008, [Online]).

Also documented in M.E. are severely reduced blood flow to the brain 72 hours post-exertion, blood pressure readings as low as 80/40 and pulses as high as 150 at rest or after a period of time being upright, and so on. It is also worth noting that these abnormalities found on testing of M.E. patients exclude the most severely affected patients, who are too ill to be subjected to such tests.

Research has also proven that how much physical and cognitive overexertion a person can tolerate without serious damage depends on the severity of their illness. We know that moderately affected patients can die from exercise sessions. For example, there is the case of the UK MP Brynmor John who had M.E. and was advised to 'exercise himself back to fitness' and who as a result of complying with this advice collapsed and died coming out of the House of Commons gym. Then there is the case of Sophia Mirza, in the UK who died from M.E. after being forced into inappropriate and abusive psychiatric care. Sophia had severe M.E. and was of course not capable of any exercise. Nonetheless, she was forcibly removed from her home and given inappropriate care. She was cruelly killed by being forced to perform what for most people would have been only very minor or trivial exertions.

Consider the fact that trauma victims are sometimes stopped from being moved (to another better equipped hospital for example), due to the fact that they are in shock and in a fragile state due to severe blood loss. How is this different to what is happening with the severe M.E. patient who only has 50% *or less* of the expected circulating blood volume? It isn't. (The trauma victim at least has most of this blood loss *replaced* by blood transfusion as soon as possible, which of course does NOT happen for the M.E. patient who must put up with this extremely low circulating blood volume for *years* at a time!) It seems clear that those with critical illnesses or injuries can indeed be significantly affected, and are at serious risk, by what would be only very minor bodily stresses to other patients. The same is true of severe M.E. patients.

It also seems clear that if those with only moderate M.E. can and do die from the illness, then those with far more severe pathology and severe disability are at increased risk. If patients exceed cardiac output by even 1%, they die. There are severe M.E. patients who are so ill and have such poor cardiac output that they must spend all day in a dark, quiet room, alone, unmoving and unthinking, and yet even this level of rest is not enough for their bodies to cope with normal bodily processes without difficulty. For these patients, even being at complete rest counts as 'overexertion' and is too great a burden for the body to manage. Clearly a trip out of the house or a brief period upright could very easily constitute the fatal 1% worth of overexertion. This is just simple logic.

Although death is a real possibility with a trip out of the house or other overexertion, most often death will not occur. A relapse is a certainty, however, if someone with severe M.E. is overexerted. This should be taken just as seriously as the possibility of death; the suffering caused by a relapse in severe M.E. patients can seem crueller than death.

While it maybe seem unkind to compare the experience of severe M.E. to death, can you imagine what it is like to be so ill and disabled, to be in what feels like 10/10 pain much of the time, and then to suddenly have your pain and suffering levels DOUBLE just because of one day's or one week's 'activity?'

Can you imagine what it's like to lose years of sacrifice and discipline, and slow improvement, hard-won through intensive rest, in just one day or one week? To have all your hard work suddenly count for absolutely nothing? To not only lose the small gains you made, but to end up even worse off than before you started?

Can you imagine being so severely ill and disabled that you have to spend 22 hours a day or more in a completely silent dark room, trying hard not to even think or move or feel very much lest you become far more ill? Imagine that all you had to look forward to, to focus on and cling to in your worst moments, was watching an hour of TV, listening to some quiet music occasionally, a half hour talking quietly to friends and family – or your own children – or using the computer (lying down) or a few minutes spent outside in the garden or playing with a beloved cat or dog.

Can you imagine suddenly losing the ability to do ANY of these things anymore, and having no respite or distraction from the agony-filled dark quiet room at all? Can you imagine this situation lasting months or even years, never knowing if things will ever return even to the level of illness prior to relapse? M.E. can be more severe and disabling than almost any other disease there is. For those of us with severe M.E., the price that we pay for 'activity' is extreme and prolonged. Severe M.E. can truly be a living death.

- Of course this is only a brief sample of some of the M.E. research available and just some of the testable abnormalities that have been documented in M.E. This is in no way an exhaustive list. See What is M.E.? and Testing for M.E. for more information.
- For more information about M.E. fatalities (and abuse leading to death) see: What is M.E.?
- Note that as well as sudden death, premature death occurs by an average of 25 years in M.E., because of the way the heart is affected, and so on. For more information see: M.E. Fatalities and The severity of M.E. on the website.
- Note that the warnings about severely affected patients needing to avoid all non-essential hospital visits do not apply to moderately affected patients, many of whom can tolerate such trips and may get extra medical care as soon as it is needed.
- Note that as well as the urgent need for at-home medical care for severe M.E. patients there is also a need for the option of home schooling for children affected by M.E.

2. How important is appropriate rest in M.E.?

Based on repeated studies of patient outcomes, M.E. patients who are given advice to rest have the best prognosis. As M.E. expert Dr Melvin Ramsay explains:

> The degree of physical incapacity varies greatly, but the [level of severity] is directly related to the length of time the patient persists in physical effort after its onset; put in another way, those patients who are given a period of enforced rest from the onset have the best prognosis. Since the limitations which the disease imposes vary considerably from case to case, the responsibility for determining these rests upon the patient. Once these are ascertained the patient is advised to fashion a pattern of living that comes well within them.

According to Dr Elizabeth Dowsett any M.E. patient can also be stopped from deteriorating further and at least stabilised (if not in time experiencing some level of improvement) through appropriate care and rest. For more information on this see: What is M.E.?

3. What does 'rest' mean exactly in this context?

Resting means completely different things at different severity levels of illness. For the mildly ill resting may mean watching TV or perhaps sitting in a chair reading a book or having a quiet night in with friends. For the severely ill, these activities are not at all restful and indeed would provoke relapses.

For the very severely ill, resting means lying down in a dark room, in silence and with no sensory input at all (such as TV or radio or light) and not moving physically or engaging in any type of cognitive activity. Clothing must also be comfortable and the room must be neither too warm nor too cold. For the very severely ill a better term would be 'complete incapacitation,' rather than 'resting.' The term 'resting' implies that the inactivity is optional and this is not the case in the severely ill who are often 'resting' (i.e. incapacitated) because it is physically impossible for them to do anything else.

For moderately ill patients resting means something somewhere between the two extremes.

Of course for the very severely ill there will be no safe or symptom-free activity limit. Concepts of pacing or of keeping activity at a level which does not cause immediate or delayed symptoms are useless. Indeed, a sizeable proportion of the very severely ill may well be so badly affected in the first place BECAUSE of overexertion in the early stages of their illness - because they were not told how important it was to rest or were not allowed to rest adequately. This is extremely common in M.E. It is a tragedy and an absolute disgrace.

Note that I have never heard of anyone with M.E. who is *too* restrictive with their activity levels; the problem is always the opposite, if anything. It is human nature to want to do things and to want to live and experience life as much as possible. It is very difficult for the person with M.E. to be unable to do so many things and it requires enormous discipline to avoid overexertion. Severe M.E. restricts life to a degree that healthy people might find hard to imagine, but patients have learnt from bitter experience many times over the extreme negative consequences of overexertion. Patients are reminded of this every week if not every day because even with careful control, limits can be misjudged or tasks can take a greater toll than expected.

For most patients it is much harder to rest adequately than it is to keep pushing yourself to do things even to the point of worsening the illness. It is often much easier emotionally to just keep doing things and suffer the dire consequences in the short- and long-term, than to stand up to extreme pressure from friends, family and medical staff for these activities to be completed as they were before the patient's illness.

Resting so endlessly for many years on end is much harder than you can imagine. It has been observed many times that learning to walk or speak again after a stroke or accident would be so much easier than having to just rest endlessly and do almost nothing and to have no distraction from the extreme pain. People with M.E. would give anything to be able to work hard to improve their illness, and to be improving every day instead of staying the same or getting worse. The problem of M.E. patients under-reporting or underestimating their ability levels just does not exist.

This is not about patients being as inactive as possible. Of course a person with moderate M.E. does not need to live with the same restrictions as someone with severe M.E. The point here is just that patients must stay within their individual post-illness limits.

Increasing the activity levels of someone with M.E. beyond their individual limits can only ever be harmful. It really doesn't matter if this is done gradually or all at once.

The evidence which shows that some 'CFS' patients are merely deconditioned and can be restored to health through graded exercise programs is based on patients who DO NOT have M.E. None of the various cardiac, cardiovascular, immunological, neurological, cognitive, muscular, and other abnormalities present in M.E. sufferers – which together cause the high level of disability associated with M.E. – can be explained by mere 'deconditioning.' *Patients who improve with graded activity programs do not have M.E.* It should go without saying that treatment of one disease cannot be determined by studying a completely different and unrelated (and mixed) patient group. Yet this essential medical and logical guideline is all too often ignored when it comes to M.E. In this case, money speaks louder than logic, science or ethics. Please don't fall for this nonsense about 'deconditioning' or about 'CFS' supposedly being just another term for M.E. It has nothing to do with M.E. For more see: Smoke and Mirrors on the website.

To summarise:

- No one with M.E. is *too* restrictive with their activity levels and M.E. patients do not underestimate their activity levels
- It is very difficult for M.E. patients to restrict their activity levels, and requires a high level of discipline
- M.E. patients know from bitter experience the negative consequences of overexertion
- The appropriate activity level depends on the severity of each patient's illness
- The symptoms of M.E. are not caused by deconditioning
- Graded exercise does not help M.E.; if a patient improves with graded activity, they do not suffer from M.E.
- Some patients that qualify for a 'CFS' diagnosis may improve with graded exercise, but these patients do not suffer from M.E.

4. What does 10/10 pain and suffering mean in this context?

A few years ago in The M.E. Ability Scale I mentioned 10/10 pain. I have experienced this on about a dozen occasions, which are burned into my memory. It is indescribable how intense the suffering and pain can be, at their worst. However it must be said that the more common 8/10 level pain is unbearable too, especially when it is very prolonged. I would also like to point out that I have severe M.E., but I am *far* from being the most severely affected.

The pain and suffering of M.E. have a number of different 'flavours.' The experience can be made up of severe nausea, vertigo and disequilibrium, cold and hot fevers or feeling both very cold and very hot at the same time, feeling 'poisoned' and very ill, pain in the glands and throat, muscle pain, twitching and uncontrollable spasms, difficulty breathing and breathlessness, cardiac pain and pressure and dysfunction that feels like a heart attack, a feeling of having a heart attack in every organ (caused by lack of blood flow to these organs), sensations of pain and terrible pressure in the brain and behind the eyes, stroke-like or coma-like episodes, abdominal pain and pain/discomfort following meals, seizures and 'sensory storms' (while conscious) and, lastly, an inability to remain conscious for more than a few minutes or hours at a time, or for more than a few hours each day in total. Any one of these problems can cause severe suffering. What makes severe M.E. so terrible is that the patient is almost always dealing with a large number of these horrific problems *all at once.*

More than 60 different symptoms of M.E. have been officially documented. Symptoms of M.E. include:

> Sore throat, chills, sweats, low body temperature, low grade fever, lymphadenopathy, muscle weakness or paralysis, muscle pain, muscle twitches or spasms, gelling of the joints, hypoglycaemia, nausea, vomiting, vertigo, chest pain, cardiac arrhythmia, resting tachycardia, orthostatic tachycardia, orthostatic fainting or faintness, circulatory problems, opthalmoplegia, eye pain, photophobia, and other visual and neurological disturbances, hyperacusis, tinnitus, gastrointestinal and digestive disturbances, allergies and sensitivities to many previously well-tolerated foods, drug sensitivities, stroke-like episodes, nystagmus, difficulty swallowing, paresthesias, polyneuropathy, myoclonus, temporal lobe and other types of seizures, an inability to maintain consciousness for more than short periods at a time, confusion, disorientation, spatial disorientation, disequilibrium, breathing difficulties, sleep disorders; sleep paralysis, fragmented sleep, difficulty initiating sleep, lack of deep-stage sleep and/or a disrupted circadian rhythm and neurocognitive dysfunction including cognitive, motor and perceptual disturbances (Bassett, 2010, [Online]).

Dr Cheney writes: '80% of cases are unable to work or attend school. We admit regularly to hospital with an inability to care for self' (Hooper et al. 2001 [Online]). M.E. has been found to be more disabling than M.S., heart disease, virtually all types of cancer, patients undergoing chemotherapy or haemodialysis. It is comparable to end-stage AIDS, i.e. to how ill and disabled an AIDS patient is *2 weeks before death (*Hooper & Marshall 2005a, [Online]). However, in M.E. this high level of suffering is not short term as it is for end-stage AIDS patients. The body has few limits on how bad pain and disability can be without actually killing the sufferer, and how long the sufferer can remain in this state. This high level of suffering can last uninterrupted for decades.

5. What is homeostasis?

Homeostasis is the ability of a living organism to regulate its internal environment to maintain a stable, constant condition, by means of multiple dynamic equilibrium adjustments, controlled by interrelated self-regulation mechanisms. Homeostasis is one of the fundamental characteristics of living things. It is the maintenance of the internal environment within tolerable limits.

M.E. causes a loss of the ability of the CNS (the brain) to adequately receive, interpret, store and recover information which would enable it to control vital body functions. There is a loss of normal internal homeostasis; the individual can no longer function systemically within normal limits.

Metabolic problems at a cellular level also contribute to this inability to maintain homeostasis in M.E. Expert Dr Byron Hyde explains: 'In MRI spectography of arm muscle of M.E. patients, it has been shown that because of an abnormal build-up of normal metabolites, the muscle cell actually shuts down to prevent cell death.' This is what is happening to the M.E. patient's cell physiology in every muscle (including the heart) and in the brain as a result of physical and cognitive activity and/or overexertion; there is 'cell field shutdown' to prevent the death of the cell. See: The importance of avoiding overexertion in M.E. for more information.

6. Is M.E. a stable illness?

One can probably observe people with some illnesses carefully for an hour or so and collect a lot of good information about what they can and can't do, how severe their illness is, and what their usual symptoms are from day to day. However M.E. is not one of those illnesses. M.E. is *not* a stable illness.

Observing the average M.E. sufferer for an hour – or even a week or more – will not give an accurate indication of their usual activity level because the severity of M.E. can wax and wane throughout the month, week, day and even hour. Also, people with M.E. can sometimes operate significantly above their actual illness level for short periods of time thanks to surges of adrenaline – albeit at the cost of severe and prolonged worsening of the illness afterward. Relapses and worsening of symptoms are also very often also significantly delayed (there may be both an acute AND a delayed reaction).

Just observing someone with M.E. do a certain task should not be taken to mean (a) that they can necessarily repeat the task anytime soon, (b) that they would have been able to do it at any other time of day, (c) that they can do the same task every hour, day or even every week, or month, or (d) that they wont be made very ill afterwards for a considerable period because they had to really push themselves to do the task.

Often a considerable rest period is needed before and after a task, which may be hours, days, weeks or months long. For example, someone may need 2 weeks rest before an outing, for example, and may then spend 3 weeks extremely ill afterwards recovering from it. Just observing them in the 2 hours they were 'out and about and mobile' is of course not at all representative of their usual ability levels.

Most importantly, because the worsening of the illness caused by overexertion may not even begin until 48 or more hours afterwards (when most observers are long gone), it's impossible to tell by seeing an M.E. patient engaged in an activity, whether that activity is so far beyond the patient's limits that it will end up causing a severe or even permanent worsening of the illness (or 'relapse'). To be blunt, the activity may even end up killing the patient. This isn't common (the death rate is estimated at 3%), but deaths can and do occur. Thus, observers who see an M.E. patient engaged in an activity have no idea what the consequences of this activity may be.

- What is an adrenaline surge? Adrenaline is often referred to as the 'fight or flight' hormone as it kicks into action in situations of potential danger. However, adrenaline also kicks in when the body is in physiological difficulty, which is very often what is happening to severe M.E. sufferers. Adrenaline surges make the heart pump faster and raise the blood pressure, forcing blood around the body with greater force to supply the muscles with more oxygen, so that they can make a greater effort. Surges of adrenaline increase the metabolism. They also relax and dilate the airways so that more oxygen than usual can be taken in. Adrenaline surges can also decrease the amount of pain felt. As a result of all of these factors, adrenaline surges – while they last – have the ability to increase physical speed, strength and other physical abilities.

  Unfortunately, when these bursts of adrenaline wear off – as they must – people with M.E. are left far more ill as a result for many days, weeks, months or even years. People with M.E. are harmed by adrenaline surges, both by the physiological stress to the body of the changes caused by adrenaline, and by the extra activity which adrenaline enables, which may be far beyond the body's normal limits so that such activity causes damage. For every short term 'gain' there is a far greater loss overall.

7. Why is M.E. not the same thing as 'CFS'?

The terminology is often used interchangeably, incorrectly and confusingly. However, the DEFINITIONS of M.E. and CFS are very different and distinct, and it is the definitions of each of these terms which are of primary importance. *The distinction must be made between terminology and definitions.*

Chronic Fatigue Syndrome is an artificial construct created in the US in 1988 for the benefit of various political and financial vested interest groups. It is a mere diagnosis of exclusion (or wastebasket diagnosis) based on the presence of gradual or acute onset fatigue lasting at least 6 months. If tests show serious abnormalities, a person no longer qualifies for the diagnosis, as 'CFS' is 'medically unexplained.' A diagnosis of 'CFS' does not mean that a person has any distinct disease (including M.E.). The patient population diagnosed with 'CFS' is made up of people with a vast array of unrelated illnesses, or with no detectable illness. According to the latest CDC estimates, 2.54% of the population qualifies for a 'CFS' (mis)diagnosis. Every diagnosis of 'CFS' can only ever be a misdiagnosis.

Myalgic Encephalomyelitis is a systemic neurological disease initiated by a viral infection. M.E. is characterised by (scientifically measurable) damage to the brain, and particularly to the brain stem which results in dysfunctions and damage to almost all vital bodily systems and a loss of normal internal homeostasis. Substantial evidence indicates that M.E. is caused by an enterovirus. The onset of M.E. is always acute and M.E. can be diagnosed within just a few weeks. M.E. is an easily recognisable distinct organic neurological disease which can be verified by objective testing. If all tests are normal, then a diagnosis of M.E. cannot be correct.

M.E. can occur in both epidemic and sporadic forms and can be extremely disabling, or sometimes fatal. M.E. is a chronic/lifelong disease that has existed for centuries. It shares similarities with M.S., Lupus and Polio. There are more than 60 different neurological, cognitive, cardiac, metabolic, immunological, and other M.E. symptoms. Fatigue is not a defining or even essential symptom of M.E. People with M.E. would give anything to be only severely 'fatigued' instead of having M.E. Far fewer than 0.5% of the population has the distinct neurological disease known since 1956 as Myalgic Encephalomyelitis.

The only course of action that makes any sense is for patients with M.E. to be studied ONLY under the name Myalgic Encephalomyelitis – and for this term to be used ONLY to refer to a 100% M.E. patient group.

The problem is not that 'CFS' patients are being mistreated as psychiatric patients; 'CFS,' as a wastebasket diagnosis, includes all sorts of fatiguing illnesses including psychiatric illnesses. 'CFS' is associated with psychiatric illness; for many patients this is inappropriate, but some patients misdiagnosed with 'CFS' actually *do* have psychological illnesses.

There is no such disease as 'CFS' – that is the entire issue. The vast majority of patients misdiagnosed with 'CFS' *do not* have M.E. The only way forward, for the benefit of all involved, is that:

1. The bogus disease category of 'CFS' must be abandoned completely.
2. The name Myalgic Encephalomyelitis must be fully restored (to the exclusion of all others) and the World Health Organization classification of M.E. (as a distinct neurological disease) must be accepted and adhered to in all official documentations and government policy.
3. The bogus disease category of 'CFS' must be abandoned (along with the use of other vague and misleading umbrella terms such as 'ME/CFS,' 'CFS/ME, ' 'ME-CFS,' 'CFIDS,' 'Myalgic Encephalopathy' and others), for the benefit of all patient groups involved. Science, logic and ethics must prevail over mere financial and political concerns.

Governments around the world are currently spending $0 a year on M.E. research. Considering the brutal severity of the illness, and the vast numbers of patients involved, ranging in age from two years to adults, this is a worldwide disgrace. How much longer will the world be fooled by the paper-thin 'CFS' scam, which has been proven to be financially motivated?' The fiction of 'CFS' represents outright medical fraud, involving serious medical abuse and neglect of patients, on a truly massive scale.

To read a fully-referenced version of the medical information in this text compiled using information from the world's leading M.E. experts, please see the 'What is M.E.?' paper on page 113 of this book or on the HFME website.

Acknowledgments
Thanks to Lesley Ben and Emma Searle for editing this paper. Thanks also to everyone who offered suggestions and comments as I was writing this paper.

Relevant quotes
'Those who are most injured or die are easily recognized at disease onset or shortly after as CNS, cardiovascular, or organ injury. Because of their overwhelming illness and the specificity of the end-organ injury, they are never diagnosed as M.E. except in epidemic or cluster situations.'
DR BYRON HYDE

'Documented deaths have occurred in several M.E. epidemics, but are best documented in the Cumberland epidemic and were well known in the Akureyri epidemic. All of these deaths involved CNS injury.'
DR BYRON HYDE

Assisting the M.E. patient in managing relapses and adrenaline surges

 Myalgic Encephalomyelitis (M.E.) patients have strict limits on how active they can be. If these limits are breached, symptoms worsen immediately and there is also a further deterioration 24 – 48 hours later, as well as the very real potential for repeated or severe overexertion to prevent significant recovery, cause disease progression or even death.

It is very important that M.E. patients stay within their limits. Unfortunately, M.E. patients may find staying within these limits all of the time very difficult for a number of reasons.

This paper explains how carers, doctors, and also friends, family members and partners of M.E. patients, can help patients to avoid overexertion, ensuring their best possible long-term health outcome. It also describes the characteristics and signs of adrenaline surges and relapses in M.E. for the benefit of these individuals, as well as for newly ill M.E. patients themselves.

This paper is designed to be read together with the more detailed <u>Hospital or carer notes for M.E.</u> paper.

What is an adrenaline surge and how does this affect M.E. patients?

People with M.E. can sometimes operate significantly above their actual illness level for certain periods of time thanks to surges of adrenaline – albeit at the cost of severe and prolonged worsening of the illness afterward.

Adrenaline is often referred to as the 'fight or flight' hormone as it kicks into action in situations of potential danger. However, adrenaline also kicks in when the body is in physiological difficulty, which is very often what is happening to severe M.E. sufferers. Adrenaline surges make the heart pump faster and raise the blood pressure, forcing blood around the body with greater force to supply the muscles with more oxygen, so that they can make a greater effort. Surges of adrenaline increase the metabolism. They also relax and dilate the airways so that more oxygen than usual can be taken in. Adrenaline surges can also decrease the amount of pain felt. As a result of all of these factors, adrenaline surges – while they last – have the ability to increase physical speed, strength and other physical abilities.

Unfortunately, when these bursts of adrenaline wear off – as they must – people with M.E. are left far more ill as a result for many days, weeks, months or even years of overexertion. People with M.E. are harmed by adrenaline surges, both by the physiological stress to the body of the changes caused by adrenaline, and by the extra activity which adrenaline enables, which may be far beyond the body's normal limits so that such activity causes damage. For every short term 'gain' there is a far greater loss overall.

Surges of adrenaline can last hours, days, weeks or even months at a time.

These adrenaline surges are a bit like owning credit cards. They allow patients to do things that they could never otherwise do, or 'afford.' But the interest rate is extortionate, sky high, a killer. Clytie, a very ill M.E. patient, explains her adrenaline surges "in terms of money." In the same way that a financially limited person could purchase a Ferrari, she says, only to be plagued by debt and potentially resorting to desperate measures to settle the score, a seriously ill M.E. patient can *overspend* in order to perform some activities. But sooner or later the loan sharks are going to show up at your door. Paying that debt, Clytie explains, "could take you a lifetime."

M.E. expert <u>Dr Melvin Ramsay</u> explains:

> The degree of physical incapacity varies greatly, but is directly related to the length of time the patient persists in physical effort after its onset; put in another way, ***those patients who are given a period of enforced rest from the onset have the best prognosis***. Since the limitations which the disease imposes vary considerably from case to case, the responsibility for determining these rests upon the patient. Once these are ascertained the patient is advised to fashion a pattern of living that comes well within them.

Why do M.E. patients sometimes overexert themselves, considering the severe consequences?

There are many reasons why this occurs, including the following:

- Once a patient gets going, stopping can be very difficult. This is due to neurological problems with stopping and starting new tasks easily, and also because once an adrenaline surge has occurred, it takes a long time to wear off.

- When a patient has become very ill from overexertion or is in the middle of an adrenaline surge, judgement can be affected, and the patient may be lost in the moment and not realize how important it is for them to stop what they are doing as soon as possible.

- Resting after a relapse is often very difficult for the M.E. patient emotionally. It can be very difficult to lie in a dark quiet room in extreme pain and worse, with no distraction from it. It can be tempting to keep the adrenaline surge going in a small way, to put off the crash. (A bit like drinking more alcohol the day after a night of heavy drinking to delay the inevitable hangover.)

- Many patients with M.E. have been treated appallingly in the earlier stages of their illness. Medical abuse is very common. Patients have often been told or forced to keep pushing through their limits and that this is what they must do if they ever want to recover. This ignorant advice has forced many patients to develop a very high tolerance for pain and discomfort – and this can be a real obstacle when it comes to training oneself to rest appropriately when experiencing minor symptoms. Patients have often become very used to paying a high price afterwards for every little bit of fun they have or every task they do, and have had to accept this as a way of life for so long that change can be difficult.

- Many patients, along with much of the general population, have a strong work ethic and at times find *not* pushing themselves to do things very difficult. M.E. patients need a level of discipline as high as the average Olympic athlete to control and restrict all their activities so completely for years.

- M.E. is an acute onset disease which means that patients go from healthy to very disabled from one day to the next. Being so disabled so suddenly is shocking and takes a long time to get used to. Particularly in the early years of M.E., patients often feel the need to constantly push at the boundaries to work out where their limits are. This is sometimes tied to denial of the realities of the disease, and a desire to keep ignoring physical limits in the madly optimistic hope that this will make them go away as quickly as they came.

Other reasons include:

- Due to the brutal severity of M.E., some M.E. patients must overexert just to live or to have a little bit of basic human contact.

- Many M.E. patients overexert themselves as they have no other choice, and do not have the appropriate financial or practical support they need due to the political situation facing M.E. patients. Patients with M.E. are also often forced to overexert themselves in order to get the welfare payments they are entitled to (and need to have to survive) and to try and get some basic medical care.

- Many M.E. patients are told that 'everyone recovers, it is just a matter of when, and at most you will be well in 5 years' by ignorant doctors, websites and patients that do not understand the massive difference between the neurological disease M.E. and mere post viral fatigue syndromes. They have no idea that there is a big long-term cost with every short-term relapse, and by the time they do get this information it is often far too late and they have become severely affected and greatly harmed their chances of future recovery.

- Many M.E. patients are told that they are not ill, and that they cannot do things simply because they believe they can't. This abusive brainwashing can lead many patients to try again and again to push past their limits in a brave but misguided effort towards a 'mind over matter' approach to the disease. Unfortunately, many patients have so much faith and trust in their doctors that they ignore what they know to be true about their own bodies and their disease for many months or even years, often ending up severely affected and disabled as a result.

What signs may indicate a relapse or an adrenaline surge?

Signs that an M.E. patient is overexerting and/or running on adrenaline may include the following:

- Very fast and continuous talking is a sure sign of an adrenaline burst. Speech may also become very loud as the patient becomes unable to modulate their volume level. This may also be accompanied by fast and jittery movements. Speech may make evident feelings of euphoria, over-excitability or wild optimism and will often be less well-considered than normal. The patient may also sit up or stand for longer than usual (without realising they are doing so) or get fired up to undertake tasks that they would usually be too ill to do. (Big cleaning or organisational tasks for example.)

- After or during an adrenaline surge, sleeping and resting is very difficult as the patient feels 'wired' and very *un*-fatigued or sleepy. Sleep onset may be delayed for many hours, perhaps leaving the patient unable to sleep. The patient may also only be able to sleep for one or two hours at a time, awaken for a few minutes many times during the night, and/or may experience very light sleep where the slightest noise wakes them up.

- Particularly in the first few years of the disease, patients may sleep, or be unconscious, for much longer than usual after overexertion, perhaps 12 – 16 hours or more. In extreme cases, the patient may be unable to maintain consciousness for more than a few hours a day.

- When suffering an acute neurological episode M.E. patients may be mistaken for being drunk or high on drugs. They may slur their words, talk very fast and ramble, seem euphoric and have very poor balance.

Other signs of an adrenaline surge include:

- A lack of facial expression and 'slack' facial muscles and/or extreme facial pallor.

- A burning sensation in the eyes and/or an inability to tolerate visual stimulus and to keep the eyes open.

- Excessive water drinking (to try to boost blood volume).

- Excessive hunger and a desire for sugar- or carbohydrate –rich foods. Even after eating, the patient may feel as if they have very low blood sugar and may need to eat far more often than usual.

- Sweating or shortness of breath after minor exertion.

- Visible shaking of the arms or legs or twitching facial muscles.

- Paralysis and weakness in the muscles or an inability to move, speak or understand speech.

- Sudden loss of ability to walk.

- Very sore throat and/or painful and tender glands in the neck (and possibly other flu-like symptoms).

- Distinctly pink, purple or blue feet or legs, with white blotches, after standing or sitting for too long.

- Patients may complain of a severe headache or feeling of pain or pressure at the base of the skull. This may also be accompanied by pain behind one or both eyes or ears, or blackouts.

- Sudden onset ringing in the ears or loss of hearing.

- During and after overexertion, a patient's pulse will often become much faster (150 bpm or more), their blood pressure will become lower and their temperature may rise and they may feel very hot (or alternating hot and cold). Pulse and/or temperature measurements may be useful in determining when a patient is overexerting. As blood pressure readings tend to be abnormal only when the patient is standing or sitting upright – which in itself causes relapse - this test will often be counter-productive and inappropriate.

Other things to be aware of:

- Sometimes when the patient is running on adrenaline, it is very obvious that this is what is happening. The patient is able to do more than usual but feels very unwell and wired; a bit like they haven't slept in days but have had a LOT of coffee (or other stimulants). At other times, particularly where the adrenaline surge is long-lasting, the adrenaline effects can be more subtle and can easily be mistaken for genuine wellbeing for a period of time.

- When a patient declares that they are improving and suddenly able to do tasks again which they have not been able to do for many months or years – and this occurs right after a very big task has been completed such as a house move or a very taxing trip to the doctor, it is almost certainly an adrenaline surge and not a real improvement. The big task was well beyond their limits and so the body has released a surge of adrenaline just to cope.

    Unfortunately, this type of lower-level but prolonged adrenaline surge will often be less easily recognised for what it is by the patient, especially where there are problems with memory and placing events which occurred weeks ago in the appropriate timescale. (Events which occurred more than a few days ago may be forgotten, or seem to have occurred much longer ago than they actually did.)

    This type of adrenaline surge can sometimes fool even the most experienced M.E. patients. Having hope of improvement replaced with the realities of a severe relapse can be very disheartening, to say the least.

- When a patient regularly pays a big price for doing small tasks but then suddenly pays only a small price for a big task, suspect an adrenaline surge. There is just no such thing as a free lunch with M.E. (When a relapse is expected and *doesn't* occur, that indicates use of the 'credit card' or that the patient's body is 'writing cheques it can't cash' as it were.)

- Often a considerable rest period is needed before and after a task, which may be hours, days, weeks or months long. For example, someone may need 2 weeks rest before an outing, and may then spend 3 weeks extremely ill afterwards recovering from it. The need for a long rest period before a task is a sign that this task is not within usual limits and will probably require an adrenaline surge to be completed and so be detrimental to the patient's long-term health.

    Ideally, a patient will only complete tasks which can be done daily or every second or third day without causing relapse. The goal is to do only 80% of the activity that can be done sustainably each day.

- Thanks to adrenaline surges, a moderately ill patient may spend several hours a day, one day a week studying or working and then 6 days extremely ill and disabled, or be able to struggle through study or work part-time and spend the rest of their time extremely ill and disabled. This type of schedule can only be kept up for a few years at best, as the patient becomes sicker and sicker and less able to bounce back from relapses.

- Some patients will manage their limits very carefully but still exhibit signs that they are running on a low level of adrenaline most or all of the time. This is likely an indication that more rest is needed and that more challenging tasks should, if possible, be scaled back or discontinued.

- When the adrenaline surge starts to wane, the patient will often feel very irritable. Part of this is due to problems with blood sugar and so eating a substantial meal can help the patient feel better both physically and emotionally. Cravings for sugar and carbohydrate-rich foods are common at this time, but a meal containing some protein, fat and some low glycaemic load carbohydrate foods is a better choice. Part of this is also an emotional response, as coming down form an adrenaline high is very difficult emotionally. A patient has just had a reminder of how it might be if they were not as ill and disabled, and also has a significant worsening of their symptoms and disability level.

    The patient may also feel very cold and shaky and even more sensitive than usual to light and noise as an adrenaline surge starts to wane.

- When a person with M.E. starts to rest after an adrenaline surge, it takes a little while for the adrenaline to wear off, so the patient will start resting and gradually begin to feel more and more unwell.

Unfortunately the patient will have to go through a period of feeling much worse, in order to feel better. Starting to feel worse shows that the adrenaline is wearing off and that the patient is resting properly. The period of feeling very ill may last for hours, days, weeks or longer, depending on how ill the patient is and how much they overexerted.

In contrast, when an M.E. patient who has not overexerted rests they will feel better right away and this improvement may continue to build over time. Thus feeling much more ill after a period of resting is another sign that the patient has been running on adrenaline and overexerting.

What can you do to help?

- When you notice fast talking, and other signs of an adrenaline burst or surge, encourage the patient to slow down. Perhaps remind them that the sooner they rest, the better off they will be.

- When a patient is talking very fast and very loudly, subtle reminders to speak more quietly may be helpful. (Remember that slow talking may in fact be a very good sign of health and of living within limits!)

- When you notice a patient sitting or standing when they do not need to, or for longer than they can usually cope with, a reminder to lie down may be helpful as the patient may not realise that they are standing up for too long.

- When, out of necessity, the patient has completed a big task that was far beyond their, be aware from the outset that this will cause a surge of adrenaline to be released. This surge will affect the patient during the event but probably also for some days, weeks or months afterwards. The patient may feel somewhat less ill and be able to do some tasks which they haven't managed for some time. But tasks done using adrenaline surges come at a very high cost long-term and so must be strongly discouraged and absolutely never encouraged.

  Making sure that the patient is aware of this characteristic of M.E. before a big task is completed - and before they misinterpret these signs of overexertion and illness as an improvement in their condition - can only be helpful.

- When a big adrenaline surge has occurred, the only way to stop it is to make sure that the body is no longer placed in physiological difficulty. For the severe M.E. patient, this will mean at least 3 days of complete rest. (The time period of rest needed will vary with how severe and prolonged the surge is and how ill the patient is.) Adrenaline will stop being released when the body is at rest and time has passed allowing the adrenaline in the system to wear off.

  Severely affected patients will need almost complete rest constantly, to avoid adrenaline surges and relapses in symptoms.

- Do not instigate conversations with a patient when they are trying hard to come down off an adrenaline surge. You may ruin hours of solid resting by asking a question that forces the patient's body to have to rely on adrenaline again in order to reply. If possible, write down any questions you have so that the patient can answer them in their own time and in the way easiest for them. When patients are severely affected and can't speak often, you may want to devise a system whereby they can reply to yes and no questions, or questions with 2 options using hand signals or printed cards.

- If possible, if you are a carer who visits the patient's home for a few hours a day, make a set time to talk to the patient so that they do not have to be 'switched on' and ready to talk the whole time you are there. Staying 'switched on' may require an adrenaline burst and leave the patient very ill afterwards even if you only spoke to them very briefly. Having to be 'switched on' in case of interaction is almost as taxing as actually talking, for many patients.

- When you speak to an M.E. patient who is very ill or relapsing, speak slowly, calmly and somewhat softly. Do not speak loudly or shout. Do not ask stressful or difficult questions when the patient is at their most ill or if it is not their 'best' time of day.

- Don't repeat things unless you are asked or indicated to do so. M.E. patients often have a significant time delay in understanding spoken words, and they may rely on a period of silence after each statement in order to understand what you have said.

- *For more information on how to appropriately treat M.E. patients and help them to avoid relapse please see the following two essential papers: Hospital or carer notes for M.E. and Why patients with severe M.E. are housebound and bedbound.*

How should you use this information?

This paper largely focuses on patients who are at the severe end of the moderate continuum and severely affected patients who are almost entirely bedbound. It will have to be adjusted somewhat for patients who are more moderately affected, or extremely severely affected.

How much assistance and guidance you give an M.E. patient to help them minimise relapses depends on several factors. These include how close you are to the patient, how receptive they are to input from you, how ill the patient is and how well they are managing their symptoms and relapses themselves.

M.E. patients understand their own limits very well almost all of the time. All they may need from you is the occasional verbal reminder to lie down or to rest.

Perhaps in most cases, just the fact that you have a desire to help and have read this paper and that you both have a solid understanding of the challenges they face and the nature of M.E. relapses will be enough. Knowledge is power.

Conclusion

It may seem obvious that M.E. patients would always do what is best for their long-term outcome, but this is not always the case. M.E. is a very difficult disease to manage. M.E. patients are very often sick to death of all the resting, caution and explanation required of them and so anything that you can do to help is very welcome. Thank you for taking the time to read this paper.

More information

- If you know someone with M.E. and want to know how to deal with it, and what you can do to help, then please read So you know someone with M.E.?
- M.E. patients and those involved in determining treatment for M.E. patients may wish to look at the Treating M.E. paper which discusses treatment for M.E. generally, and also treatments which may support normal adrenal function. (This includes extra B complex and vitamin B5, high-dose vitamin C, vitamin A, low dose cortisol and unrefined sea salt.)
- For tips for M.E. patients on avoiding overexertion see: A quick start guide to treating and improving M.E. with aggressive rest therapy, diet, toxic chemical avoidance, medications, supplements and vitamins plus the important new paper: Tips on resting for M.E. patients.
- All M.E. patients need to have their cortisol levels checked regularly. Low cortisol levels are well documented in M.E. In some cases testing may indicate very low cortisol levels and a prescription for low dose cortisone may be required. See Testing for M.E. for more information.

To read a fully-referenced version of the medical information in this text compiled using information from the world's leading M.E. experts, please see the 'What is M.E.?' paper on page 113 of this book or on the HFME website.

Acknowledgments
Thanks to Caroline Gilliford for editing this paper. Thank you to Victoria for suggesting the topic of this paper. Thank you to Victoria, Frir, Clytie and everyone else who contributed to this paper.

References note

The foundations of the pathology and symptomatology described in this text are well documented. For referenced information on the importance of avoiding overexertion in M.E., cardiac insufficiency in M.E., deaths in M.E. patients caused by overexertion, circulating blood volume being reduced to 50% or less and very low blood pressure readings in M.E., severely reduced cortisol levels in M.E. and the delayed effects of overexertion in M.E. etc. please see: What is M.E.? Extra extended version, Testing for M.E. and The effects of CBT and GET on patients with M.E.

What is not as well documented, however, is the exact nature of the relapses and adrenaline surges in M.E. The details on adrenaline surges included in this paper have been taken largely from hundreds of patient accounts shared with me both privately and in various online groups over the last 10 years or so, as well as my own experiences as a long-term M.E. patient. Further comments and suggestions from knowledgeable patients or doctors are always welcome.

Relevant quotes

'M.E. is a systemic disease (initiated by a virus infection) with multi system involvement characterised by central nervous system dysfunction which causes a breakdown in bodily homoeostasis (The brain can no longer receive, store or act upon information which enables it to control vital body functions, cognitive, hormonal, cardiovascular, autonomic and sensory nerve communication, digestive, visual auditory balance, appreciation of space, shape etc). It has an UNIQUE Neuro-hormonal profile.'
DR ELIZABETH DOWSETT

'There is ample evidence that M.E. Is primarily a neurological illness. It is classified as such under the WHO international classification of diseases (ICD 10, 1992) although non neurological complications affecting the liver, cardiac and skeletal muscle, endocrine and lymphoid tissues are also recognised. Apart from secondary infection, the commonest causes of relapse in this illness are physical or mental over exertion.'
DR ELIZABETH DOWSETT

'This illness is distinguished from a variety of other post-viral states by a unique clinical and epidemiological pattern characteristic of enteroviral infection. Prompt recognition and advice to avoid over-exertion is mandatory.'
DR MELVIN RAMSAY & DR ELIZABETH DOWSETT

'[Legitimate descriptions of the illness are] a far cry from the hopelessly inadequate description of M.E. as 'chronic fatigue.' The distinction between fatigue and M.E. needs emphasising. If you are tired all the time, you do not have M.E. If you are feeling drained following a viral illness but are recovering over weeks or months, you do not have M.E. A central problem is the word 'fatigue' which doesn't come close to describing how sufferers can feel – comatose might be better.

Like most people with M.E. I have acquaintances who say, 'Oh I feel tired at 4pm too, and would love a snooze.' But that's not it. Minds and bodies do not function. This is nothing like fatigue.'
LYNN MICHELL IN 'SHATTERED: LIFE WITH M.E.' P 6

'If patients draw down their lifestyle to live within the means of the reduced cardiac output, then progression into congestive cardiac failure (CCF) is slowed down, but if things continue to progress, a point will be reached where there is no adequate cardiac output, and dyspnoea will develop, with ankle oedema and other signs of congestive cardiac failure.

In order to stay relatively stable, it is essential for the patient not to create metabolic demand that the low cardiac output cannot match.'
DR PAUL CHENEY [VIDEO LECTURE]

'Dr Paul Cheney explained how the bodies of patients are choosing between lower energy and life, or higher energy and death. On a physiological level, patients live in a near-death suspension, making patients feel much like they are dying, not tired.'
PEGGY MUNSON 2003

'There is a difference between diastolic dysfunction and diastolic failure: in diastolic dysfunction there is a

filling problem but the body is compensating for it and achieving enough cardiac output to match metabolic demand. Diastolic failure begins when the body can no longer compensate and there is a reduction in cardiac output. This is seen in 80% of patients. In order to stay relatively stable and avoid heart failure, it is essential for the patient not to create metabolic demand that the low cardiac output cannot match.'
DR PAUL CHENEY [VIDEO LECTURE]

'Patients have a high heart rate but a low cardiac output. There is a cardiac dimension that is independent of (but not excluding) autonomic function or blood volume. It's hard to talk about a low cardiac output without talking about the involvement of the brain and the adrenal glands. A mismatch between metabolic demand and cardiac output, even very briefly, will kill. If the cardiac output goes down, in order not to die, there is a rise in noradrenergic tone (also involving the adrenal glands) to bring the output back up. This is a serious problem, because when the adrenals are exhausted, there will be low cardiac output. There is no such thing as an [M.E.] patient who is NOT hypothyroid: this has nothing to do with thyroid failure, but everything to do with matching metabolic demand and cardiac output.'
DR PAUL CHENEY [VIDEO LECTURE]

'Order of sacrifice in cases of declining microcirculation; First is the skin; second is the muscles and joints; third is the liver and gut (patients can usually only tolerate a few foods); fourth is the brain; fifth is the heart; sixth is the lung and lastly is the kidney.'
DR PAUL CHENEY [VIDEO LECTURE]

'Among the major causes of death in [M.E.] is heart failure: 20% die of heart failure. There are two types of heart failure: systolic (which is a failure to eject) and diastolic (which is not a failure to eject, but a failure to fill properly). There are two types of diastolic heart failure: primary relaxation deficit giving rise to decreased cellular energy as seen in [M.E.] and secondary relaxation deficit as seen in hypertension, diabetes and the elderly over age 75. Primary relaxation deficit is a disorder that seems to have gone right under the radar of most cardiologists (who focus on the secondary relaxation deficit). Diastolic heart failure was first described in the 1980s but there was no significant literature until the 1990s, and no significant way to measure it until 2001. One is just as likely to die of diastolic heart failure as from systolic heart failure.'
DR PAUL CHENEY [VIDEO LECTURE]

'If your illness is M.E., the main thing you can do to help yourself is not push beyond your limits. I seriously damaged my health by pushing myself to continue at work after I became ill. I bitterly regret that now. I wish I had had access to Jodi's Hummingbird website at that time, which gives the all-important message that we must not push beyond our limits.'
LESLEY, M.E. PATIENT

'ME "old-timers" all say the same thing. Please take really good care of yourself, and don't get into the mindset of "needing" to push yourself because of things you want to do in the future.

When you have M.E., it's important to NOT push yourself, so you'll still be able to do those things in the future! You deserve to take really, really good care of yourself, especially right now, in the beginning.

I did the same thing too (pushing through, and over-exerting), and I regret every moment of it. If you are still within the first few years, and it IS really M.E., please slow down the pace of your life - you still have a chance of a meaningful recovery if you're very careful. I wish I could go back and do it all over, but since I can't, I wanted to tell you that you can avoid the mistake we've made. I am now totally disabled, and during my "healthy times", I am doing good to leave the house for a couple hours 2 times a week (and there are a LOT of M.E. patients much worse off than me!).
I also go through periods when I deal with a worsening of my symptoms (like now) and these periods are especially difficult to deal with (bed bound, house bound, etc for extended periods).'
SARAH, M.E. PATIENT

'i feel sick,foggy, achy, weak, dizzy, jangly but not yummy old fatigue: like you get when you have walked to the beach, or dug a garden bed, or shopped til you dropped..'
BARBARA, M.E. SUFFERER

'It is as if someone has frayed the ends of every nerve in the body and left them raw and exposed. It brings an overwhelming need to close down sensory input and, for many, to retreat from everyday ordinary stressors - conversation, noise, light, movement, TV - since they are agonising to deal with. Everyone said that they were not fatigued.'
LYNN MICHELL, DISCUSSES HER M.E. PATIENT INTERVIEWS FOR HER BOOK (P.24)

'i try to xplain this one in terms of money
ok, yoiu could buy a ferrari, coulnd't you?
no, i don't have the money
oh, but you could borrow lots of money, sell your house, talke up dealing drugs, gamble, and in the end
you'd have your ferrari
<blank look>
maybe for a day or so, before the l,oan sharks shoot your kneecaps off, and the gangs blackmail you, and you
have to find a way to rerpay all that money... could take you a lifetime
<rather frightened look>
yeah. yioyu get the point'
CLYTIE, M.E. PATIENT

'My worst acceleration in symptoms was when I was trying to work and go to university at the same time. I
really wish I had listened to my body and stopped earlier (although I know this is really hard to do). I might
have been able to go back and work part-time or something. As it is, I'm pretty sure the damage is permanent
now. I don't think I will ever be able to work again. On the other hand I am so glad that I did not keep going.
I'm sure that I would be a lot worse (scary thought!) if I had.'
N, M.E. PATIENT

'If it is M.E. & you continue to overdo, you may well end up sooooooo much worse than you are now. It
happened to Jodi, it happened to me, this determination that we are suffused with in our culture of soldiering
through & mind over matter & good people get well...& then we do "fight the good fight"...& with this
disease the price is very, very high & can be permanent. Hate to be a harbinger of doom, I know it is hard to
imagine being more ill, but you do have an opportunity here to avoid our fate. Hard choices I know.'
AYLWIN CATCHPOLE, M.E. PATIENT

'I am a ghost in the land of the living - forgotten, ignored and drifting on the edges of life, whispering my
message in the ears of the lucky ones who can participate in life and community. But they don't hear me.
And mine is all too often the fate of those of us existing with a disabling chronic illness.
    I have M.E., or Myalgic Encephalomyelitis, that most ridiculed entity, downgraded to something called
Chronic Fatigue Syndrome by most of the medical profession.
    I call it paralysis, muscle and cardiac failure, brain injury, a living plague that kills only slowly but does
kill, that has planted me on the sidelines of life, incapacitated and waiting for the Telethon, Walkathon,
ANYthing-a-Thon in recognition of this insidious and infectious plague that has rendered millions
worldwide house and bed-bound. I get so jealous when I see the pink-clad hordes out supporting others, who
already have better support than I can ever hope for.'
AYLWIN CATCHPOLE, M.E. PATIENT

A dedication
This paper is dedicated to my dear friend Aylwin (Jennifer) Catchpole. Aylwin was the first fellow M.E.
patient I met that really understood that the 'adrenaline surge' was a part of M.E. and was suffering with the
problems associated with it as much as I was.

Over many (short) emailed conversations over a period of years she helped me more fully understand this
phenomenon, and so very much was a contributor to this paper. Aylwin died in 2010 (before this paper was
completed), but I am sure she would have been very happy to know that this paper had been written and that
hopefully many M.E. patients will get this information at the *start* of their disease when it can do the most
good, rather than many years or decades in – like most patients do, and like we both did, unfortunately.

For more information on Aylwin see the Aylwin Catchpole memorial page on the HFME website.

Assisting the M.E. patient in managing relapses and adrenaline surges: Summary

COPYRIGHT © JODI BASSETT OCTOBER 2010. FROM WWW.HFME.ORG

 It is very important that Myalgic Encephalomyelitis (M.E.) patients stay within their limits in order to prevent relapse and disease progression and so that chances for significant recovery are not destroyed. This paper explains how carers and loved ones of M.E. patients can help patients to avoid overexertion and so have their best possible long-term health outcome.

People with M.E. can sometimes operate significantly above their actual illness level for certain periods of time thanks to surges of adrenaline released when the body is put in physiological difficulty – albeit at the cost of severe and prolonged worsening of the illness afterward. These adrenaline surges are a bit like credit cards. They allow patients to do things that they could never otherwise do, or 'afford.' But the interest rate is extortionate.

Signs that an M.E. patient is overexerting or running on adrenaline, may include the following:

- Very fast, loud and continuous talking is a sure sign of an adrenaline burst. The patient may also sit up or stand for longer than usual (without realising they are doing so) or get 'hyper' and fired up to undertake tasks that they would usually be too ill to do. Sleeping and resting is very difficult as the patient feels 'wired' and very 'unfatigued.'

- A lack of facial expression and 'slack' facial muscles and/or extreme facial pallor.

- A burning sensation in the eyes and/or an inability to tolerate visual stimulus and to keep the eyes open.

- Excessive water drinking (to try and boost blood volume) and excessive hunger and a desire for sugar- or carbohydrate –rich foods. Even after eating, the patient may feel as if they have very low blood sugar and may need to eat far more often than usual.

- Sweating or shortness of breath after minor exertion or a sudden loss of the ability to walk.

- Visible shaking of the arms or legs or twitching facial muscles.

- Paralysis and weakness in the muscles or an inability to move, speak or understand speech.

- Distinctly pink, purple or blue feet or legs, with white blotches, after standing or sitting for too long.

- Patients may complain of a severe headache or feeling of pain or pressure at the base of the skull. This may also be accompanied by pain behind one or both eyes or ears, or blackouts.

- Sudden onset ringing in the ears or loss of hearing or sore throat and painful glands in the neck.

- During and after overexertion, a patient's pulse will often become much faster (150 bpm or more), their blood pressure will become lower, their temperature may rise and they may feel very hot.

When a patient declares that they are improving and suddenly able to do tasks again which they have not been able to do for many months or years – and this occurs right after a very big task has been completed such as a house move or a very taxing trip to the doctor, it is almost certainly an adrenaline surge and not a real improvement. Improvements just do not occur after overexertion in M.E. this way. The big task was well beyond their limits and so the body has released a surge of adrenaline just to cope. Unfortunately, this type of lower-level but prolonged adrenaline surge will often be less easily recognised for what it is by the patient.

Tasks done using adrenaline surges come at a very high cost long-term and so must be strongly discouraged and absolutely never *en*couraged. You might gently remind the patient to lie down and rest if they sit up for much longer than usual, or are talking very fast and far more than usual.

Do not instigate conversations with a patient when they are trying hard to come down off an adrenaline surge as this can undo hours of resting. If possible, make a set time to talk to the patient so that they do not have to be 'switched on' and potentially ready to talk for hours at a time, as this is almost as taxing as actually talking for many patients.

The only way to stop an adrenaline surge is to make sure that the body is no longer placed in physiological difficulty. This often means at least 3 days of absolute rest. While some of the effects of overexertion are immediate, there are also secondary relapses that are delayed by 24 – 72 hours.

Ideally, a patient will live long-term only completing tasks which can be done daily or every second or third day without causing relapse.

Perhaps in most cases, just the fact that you have a desire to help and have read this paper and that you both have a solid understanding of the challenges they face and the nature of M.E. relapses will be enough. Knowledge is power.

Thank you for taking the time to read this paper. Please see the full-length version of the text for more information on adrenaline surges and M.E.

Assisting the M.E. patient in having blood taken for testing

COPYRIGHT © JODI BASSETT OCTOBER 2010. UPDATED JUNE 2011. FROM WWW.HFME.ORG

Myalgic Encephalomyelitis (M.E.) patients may experience significant difficulties in having blood taken. Problems may be caused by low blood pressure, blood clotting abnormalities (blood that clots too quickly), reduced circulating blood volume and an inability to maintain an upright posture.

These problems may be minimised in the following ways:

Pre-blood test tips

- M.E. patients should be advised to drink 1 – 2 glasses of water before blood tests, to help boost blood volume and raise blood pressure slightly. It may also help to improve hypercoagulation of the blood. A glass or two of an electrolyte drink will be even more effective than plain water. (See <u>Treating M.E.: The basics</u> for a simple recipe. At a pinch, some patients use a product similar to 'Gatorade' although these products are unfortunately very high in sugar and relatively low in salt and potassium. An unsweetened bottle of organic coconut water may be a better substitution, if available.)

- Very severely ill M.E. patients may benefit from a saline IV being given to them the day of, or the day before, blood is taken for testing, or both. The IV will help to reduce the relapse caused by the interaction with medical staff and also slightly improve circulating blood volume and blood pressure and make the task of collecting the blood samples easier also.

- As much as possible, M.E. patients should be given blood tests at their best time of day and always while lying down. This will minimise the detrimental effect on severely ill patients and also make the task of collecting the blood samples easier.

- M.E. patients that have deep veins or very thin veins that are difficult to take blood from should be advised to apply a heat pack to the appropriate arm (or both arms) for 10 – 25 minutes before blood is taken. The heat dilates the blood vessels and makes the veins easier to see and to take blood from. If veins are very difficult to take blood from, having a 15 minute warm bath before blood is taken can help, although this isn't always practical.

Blood test tips

- The medical staff collecting the blood samples should always be advised beforehand that the patient has low blood pressure and low blood volume and/or blood that clots easily.

- The medical staff collecting the blood samples should always listen to the patient or their carer about which vein or veins are the best to use. (Often there will be a significant difference in one arm being used over another.)

- Severely ill M.E. patients should always be given blood tests in their own home, wherever possible. Even short trips out of the house can cause severe relapses which can last days, weeks, months or longer.

- As puncture sites from blood draws will take longer to heal in M.E. patients, medical staff should be advised to use a different vein if possible, where tests are repeated within a short period of time.

- Some M.E. patients will bruise very easily and may feel pain from blood draws (or minimal movement or flexing of limbs) more than might be expected in other patients. Where appropriate, medical staff should be advised to be very gentle in handling the arms of M.E. patients. Younger patients with M.E. in particular may also benefit from the use of anaesthetic cream spread on the puncture site some time before blood is taken. Some adults may also prefer that anaesthetic cream be used. (This may reduce pain and also bruising and swelling at the site and so speed up the healing process.)

- Light pressure should be applied for two minutes to all puncture sites directly after the needle is removed. This helps to prevent bruising and applies to all types of needle puncture sites whether they were successful or not.

- If children (or adults) are afraid or uncomfortable around needles, slowly counting as blood is taken can help.

- Patients or carers may want to advise medical staff to use smaller bore needles in M.E. patients due to the fragility of the veins which often occurs.

- When many vials of blood are taken at one time, very severely affected M.E. patients may feel weak, dizzy or faint. This may be due to severely reduced circulating blood volume or the patient being too ill to cope with the interaction between themselves and medical staff over such a period of time – or a combination of the two. Extra electrolyte drinks are essential for these patients.

Unfortunately, even after following some or all of these guidelines, blood flow into the tube can stop after less than a minute in some severely affected M.E. patients due to very low blood pressure and reduced circulating blood volume.

Blood test aftercare tips

- Many people are sensitive to the wound dressing and if this is the case a hypoallergenic dressing should be used. Micropore tape (or even sellotape) and cotton wool are preferable to the more typically used coloured sticking tape.

Other problems to be aware of with testing M.E. patients

- *Fasting and glucose solutions:* M.E. patients will often be unable to fast for tests or drink glucose containing solutions without it making them very ill for some time afterwards. Tests which require fasting or ingestion of a glucose solution should not be given to M.E. patients unless there is a real need that justifies the relapse it will cause, because this relapse may last for some time.

- *Blood pressure monitors:* M.E. patients will sometimes have such low blood pressure that it comes up as an error message on some blood pressure monitors. If this occurs a different type of monitor should be used, if possible, and the test redone.

- *Intravenous medications:* M.E. patients often need smaller bore needles for use with IVs. Very chemically sensitive patients may also benefit from the use of glass IV flasks. As with blood draws, patients may benefit from careful handling of the arm and the use of anaesthetic creams.

- *Tests which involve standing upright for periods of time:* Standing upright for even a few minutes can be too much for many M.E. patients and can cause a severe worsening of symptoms or even a severe cardiac event. When the M.E. patient says that they have to lie down right away, this request should NEVER be ignored. See Testing for M.E. for more information. For more information on M.E. treatment, electrolyte drinks and supplements which help to thin the blood see: Treating M.E.: The basics

Acknowledgments
Thanks to Claire Bassett for editing this paper. Thank you to Nakomis, Judy, Pam, Victoria, Jo, Charmion and everyone in the HFME M.E. chat groups that contributed to this paper.

Assisting the M.E. patient in managing bathing and hair-care tasks

COPYRIGHT © JODI BASSETT AUGUST 2010. UPDATED JUNE 2011. FROM WWW.HFME.ORG

 Bathing can at times be very difficult for patients with Myalgic Encephalomyelitis (M.E.) Problems faced by the M.E. patient can include the following:

1. Problems tolerating very warm temperatures and problems tolerating very cool temperatures, or sudden changes in temperature, and difficulty warming up once a certain level of coldness is reached. Becoming very cold or hot can cause relapse.

2. Problems being upright for more than very short periods of time, or at all (e.g. standing, sitting or even lifting the head from the bed).

3. Problems with balance, vertigo, dizziness and faintness.

4. Poor flexibility, grip and muscle strength.

5. Poor muscle strength in regard to completing repetitive tasks. (e.g. tooth brushing or lifting body weight in and out of the bath).

6. The recovery period after certain tasks, or after a period of overexertion is often days or weeks or longer. (It may take longer than a day for a patient to recover from having a shower or a bath and so this task may not be able to be done daily.)

7. Problems tolerating many commonly used chemicals in personal care or cleaning products.

8. Problems tolerating sensory inputs such as noise, light, movement and touch.

9. Problems completing complex tasks or small tasks with many individual steps.

10. Ability to do tasks changes markedly with time of day and from one day to the next; this can depend on how active the patient has been in the hours, days, weeks and months previous.

To complete these tasks, special equipment or other items may need to be purchased, minor or major assistance from a carer may be necessary and/or tasks may need to be modified in certain ways or done less often. The following suggestions and tips come from people who have been ill with M.E. for a considerable time and who would like to share some of the solutions they have found with their fellow patients, in the hopes that others may benefit from their years of experience and (many) experiments. Some tips are aimed at M.E. patients themselves, while others provide information appropriate for carers and parents of M.E. kids.

Tips on showering

- A shower chair can be useful where the patient is shaky and has a fear of falling while in the shower. It can be used everyday or just on those days where it is needed. Shower chairs can be freestanding or wall mounted.

- Where standing can only be tolerated for a few minutes at a time a patient may wish to have a 1 minute shower as follows; turn on the water and while it's heating up, get undressed; step into the shower and spend about 20 seconds washing your face and then your underarms and crotch with some liquid soap; get out of the shower, quickly dry off and get dressed in clothes you've brought with you into the bathroom.

- Some patients find it easier to lie or sit on the floor of the shower, and use the shower hose to bathe.

- A warm bathroom can make undressing for a bath or shower easier. (Patients that become too cold, even for just a few minutes, may be unable to become warm again for many hours afterwards and this may trigger a relapse.)

- Some patients cannot tolerate showers and need baths as they find they need the immersion in heat to be able to bathe. Others, who cannot shower due to the need to be upright for too long, can bathe lying down in a bath. A patient's tolerance for baths or showers can also change over time.

- Non-slip shower mats, non-slip flooring in the bathroom and shower rails may be helpful for some patients.

- Patients that feel painfully cold after a shower may benefit from using an electric blanket to warm their bed before they shower.

    Where Electromagnetic Field Sensitivity is severe, the blanket can be put under the quilt, tuned on for a few hours and then turned off and pulled out of the bed before the patient gets in bed. (This is easier where the bed slept in at night is different to the bed used in the daytime.)

Tips on having a bath

- A very hot bath can cause dizziness, heart-pounding and weakness. It is not recommended that patients ever just get into a very hot bath as a healthy person might.

- Some patients that have problems tolerating the heat of baths may well be able to tolerate a warm bath if the bath is done a certain way which minimises the stress on the body and only exposes the body to the heat very gradually. Instead of just getting into a warm bath and shocking the body with a huge temperature change all at once, using the following steps may help: Get into the bath at the stage that there is only 1 inch or 2 cm of warm water in it. Then fill the bath slowly (at the normal speed) with water just a bit over lukewarm or as warm as one needs to not feel shivery. Then, when bath is about 3/4 filled or a bit more, put straight hot water in until it's nice and warm. If possible, mix the water with the legs to even up the water temperatures. It is also very important to have only half (or a bit over half) the ribcage submerged, and never the full chest so that the warmth from the bath isn't too much for the body to cope with.

    A half-filled bath with moderately warm water may be better tolerated than a full bath with luke-warm water – the amount of water in the bath and how quickly you're immersed in the water make a difference too. Legs and arms may be prone to poor circulation and cold and so need warmer bath temperatures, and have a higher tolerance for them, than the torso (including the heart) and the head. Having only 50% of ones body exposed to the heat of the water makes it easier for the body to cope with.

    To make the bath last longer, let some water drain out and replace it with some more hot water when the water gets too cold.

    It is also important not to leave the bath while the water is still quite warm or too cold. Leaving when the water is too warm can leave ones heart pounding and overstimulated, while leaving when the water is too cold can cause a chill and an inability to get warm again.

    These precautions can allow even those with very severe heat sensitivity to take pain-relieving and warming baths, occasionally or even daily, with no problems. Baths should always be taken at the time of day when the patient is least ill and more able to tolerate temperature changes.

- Soaking in the bath for a time is recommended as this allows time for recovery from the effort of getting undressed and getting in. Resting or perhaps even reading a book or listening to music or an audio book for a while is also recommended so that the pain relief of the bath can be enjoyed fully. Exfoliation and the use of soaps should be left until just before the end of the bath so that the patient is only soaking in 'clean' water.

- If bath water temperature needs to be within a narrow range to avoid problems, buy a bath thermometer.

- In summer the bath might be kept filled with cold water and occasional dips or splashes taken throughout the day.

- Where strong scents can be tolerated, a few drops of essential oil added to the bath can be very pleasant, but a cheaper way to go about it can be to put two drops of oil, always combined with two or three drops of carrier oil, on the chest (right under the nose so that it can easily be breathed in). Only about a fifth as much essential oil is needed this way, and the bath doesn't become hard to clean with essential oil residue. Try ginger, lavender or rose geranium oils.

- When the patient gets so seriously cold that only a warm bath can help, but the patient isn't well enough for a bath, sometimes having a plastic bucket of warm water to soak the feet in (while lying in bed, with legs hanging off the side) for 10 minutes can help, as can a hot water bottle. These warming up ideas can also be used to prepare the patient to be warm enough to undress for a shower.

- A bath with in-built jets or a spa bath mat put in a normal bath can be very soothing, if one can afford them and put up with the noise. They avoid the chemical/mould problems common with outdoor spas.

- Where the difficulty in bathing is being able to lift ones body weight into and out of the bath, a walk-in (and sit down) bath may be an option, albeit a very expensive one. There are also bath lifts, inflatable bath lifts, swivel bath seats and hydraulic/powdered seat baths which lower one into the bath automatically, in a seated position.

- Keep a dry handtowel draped over the side of the bath that can be gripped more easily when the patient is getting in and out of the bath. This towel can also be used to dry hands before reading a book in the bath. Make sure that there is a non-slip mat for the patient to step on when getting out of the bath.

Tips on hair washing

- Patients may want to wash hair quickly while leaning over the bathroom or laundry sink. Hair should be washed only at the time of day that cool temperatures can be best tolerated so that cold water straight from the tap can be used. A pump-pack of shampoo should be placed near the sink, ready to be used. This is also the best way to wash hair where there is a need to keep the ears completely dry.

- Patients may find it easiest to wash hair sitting in a chair and with the head laid back on the sink (as at a hair salon) with help from a carer. The back of the chair should be covered with plastic and a folded towel placed around the neck and also on the edge of the sink to rest the patients head and neck on.

- Patients should make sure all soaps and shampoos and conditioners are pump packs so that valuable time and effort isn't used opening lids, turning bottles upside down and shaking them to get the product to come out and so on.

- Where hair really needs a wash but there is no time or ability to do so, consider just washing the front of the hair (the fringe) under the tap quickly.

- Hair can be washed (with the help of a carer) by swivelling your body until your head is leaning over one side of the bed over a bucket of water (doing this lying on your stomach may stress the neck less), or by using an inflatable hair washing basin for the bedbound.

- For use in emergencies, powdered shampoos may be useful. Some patients use oatmeal, cornstarch or talc powder).

- Dry hair while lying flat in bed to avoid overexertion. If hair is thick and the weather is cold, hair can be dried by lying on some pillows on the floor and laying the hair next to an electric heater for half an hour. Longer hair that is wet can be done into 2 – 4 rough plaits to let air circulate around the scalp a bit more if the patient starts to feel a serious chill.

- Micro-fibre towels are very light and small and dry hair quickly and thoroughly. On hot days, wet hair quickly under the tap and let dry naturally after towelling off briefly, to cool the whole body down.

- Hair can be washed while lying down in the bath. Perhaps where a full hair wash is not possible, the hair could just be given a water rinse between washes to help keep hair fresh longer.

- Once hair is dry, comb hair while lying down in bed using a big (and light) brush if possible, to minimise the number of strokes needed. Hair can also be brushed by a carer while the patient lies in bed.

Tips on hair maintenance

- Many patients with severe M.E. choose to cut their hair very short, for easier maintenance. This style means that there is no need for a hairdresser as hair can be cut with clippers, there is no need for hair styling before going out or having visitors, and hair is very easy to wash and also dries very quickly.

- Some M.E. patients prefer to keep their hair just long enough to tie back so that it can be easily pulled away from the face (and so easily styled for trips out of the house too). Hair on the face may make the skin on the face greasy and cause pimples, especially where hair washing isn't done as often as it might be. It can also cause problems with concentration where the M.E. patient must have an entirely distraction free environment to do simple tasks such as walking. Hair this length can be quickly styled by twisting it into a bun and securing with a hair claw. Hair claws pull at the hair less than (even snagless) elastic hair bands.

- Hair that is around shoulder length, and that is usually tied up in a bun with a hair claw or put in a ponytail can be trimmed easily by the patient either in a 2 minute standing-in-the-bathroom haircut, or in bed using a mirror placed by the side of the bed. Use a hair claw to tie back the back half of your hair and cut the front part first, then let all your hair down and brush it, then cut the other half. Tie hair into a ponytail to check for any missed longer hairs and trim any that you see.

  Or, for a quicker if messier cut, just tie hair into a ponytail and chop off all hair longer than a hands-length from the hair tie. Another technique used by some M.E. patients is to position a beanie on your head and to then cut around it, perhaps using pinking shears to give a more interesting cut. The cut can be altered by parting the hair in different ways before cutting.

  To cut a fringe/bangs which feathers hair (especially straight hair) around the face giving you a softer look, one patient recommends that you wet your hair and plaster it to your forehead then cut little Vs into the line of the fringe (preferably using pinking shears).

- Mobile hairdressers may be available in your area and may be a good idea, if you can afford them and if you can sit up for the (very long for some) 10 minutes or more required. (Many M.E. patients would find sitting up so long impossible.) Some mobile hairdressers may also cut your hair for you while you lie in bed.

- 'Bad' hair or very short hair can be covered up with a bandana, scarf or hat where desired.

- To shave legs, use an electric razor or epilator while lying down in bed, using a bath towel laid under you to catch the hair. If you have baths, you can also shave armpits and legs while lying down, just using a disposable razor. Don't waste time upright shaving legs standing up in the shower!

- Many patients with severe M.E. choose to keep their pubic hair trimmed to a very short length (with scissors or electric clippers) for reasons of improved hygiene and comfort. When using clippers, narrow clippers designed specifically for use on pubic hair may be best and your grip on them can be improved by wrapping a few rubber bands around the clippers. Clippers may be easier to use, but for some the noise of them and the effort needed to set them up may make scissors a better option.

- To shape eyebrows with tweezers in bed, use a small mirror with at least 2 x magnification. Don't waste upright time doing this standing up. If you're able to sit up, prop a mirror on a lap desk or bed tray.

Tips on oral care

- Get an electric toothbrush. (A cheap model is fine and you may want to look for cheap replacement head packs online.)

- Brush teeth in the bath, lying down, just before getting out. Use tap water to rinse out your mouth, and then spit it into the bath (away from yourself!) just before you get out.

- Patients should buy toothpaste that has a flip-top lid only, no screw tops. When tooth brushing is not possible, floss teeth and use mouthwash while lying down in bed.

- Nursing homes and hospitals use products a bit like a large cotton bud or a small sponge on a stick to gently clean patients' teeth or apply mouthwash.

- Patients that have severe jaw pain and cannot open their mouths wide enough for teeth brushing may use a finger with some toothpaste on it to clean their teeth or, even better, a slip-on textured finger toothbrush. To find these products, search for 'silicon finger toothbrush.'

Tips on hand washing

- Liquid soaps save valuable time compared to bar soaps and they make lather far more quickly (and are also less tough on painful hands and body parts).

- Some patients keep unscented baby wipes or a bottle of inexpensive vodka (!) or isopropyl alcohol next to the bed to use for hand washing so they don't have to get up to wash their hands. Other patients find these products too harsh and prefer to keep a small basin of water, washcloth and some simple liquid or bar soap by the bed for this purpose. Antibacterial products of any kind are not recommended as they contain unsafe chemicals. (All products containing triclosan should also be avoided.)

General tips

- Buy a brush on a long handle to scrub your back, or a textured back scrubbing cloth to use in the bath. Scrubbing your back now and then is important if you spend a lot of time in bed.

- An exfoliating bath mitt or loofah can make bathing quicker and more thorough. A face scrub product may also be useful. Bedbound M.E. patients tend to need to exfoliate more often than healthy people. It is especially important to exfoliate before shaving to prevent itching and ingrown hairs. A foot file or pumice stone is also recommended to keep feet soft.

- A towelling robe (or just towels) can be worn after a bath or shower for 5 minutes or more, to dry off in.

- Trim nails after bathing when they are softer.

- Covering the whole body with thick white soap suds every day, despite what some advertisers would have us believe, is not only unnecessary but is very bad for your skin! Soaps containing harsh cleansers (such as Sodium Lauryl Sulfate) are not good for your skin, and mild soap is needed only on the face, armpits and crotch. M.E. patients need only very mild plant-based products or home-made/pure products if sensitivities are severe. (Some patients also report needing to change products often due to sensitivities developing.)

- Some patients find it helpful to have faucet levers attached instead of turn taps. Levers which set a temperature (like those used in hospitals) can also be useful. Some patients will also need easier to use taps, or for taps to be lowered to be more reachable.

- If you have M.E. and bathe alone, always keep a torch handy in case the power goes off (if like most M.E. patients you have no sense of up and down at all without light).

- Bamboo towels are very soft and may be a good choice for the very severely affected M.E. patient.

- Other problems commonly faced by M.E. patients to be aware of include dry skin and many different skin conditions, scalp infections and itching, night sweats, ingrown toenails and nail infections.

Tips for when bathing isn't everyday, or is more a sponge bath than a full shower or bath:

- Keep a natural rosewater spray by your bed to spray onto face and legs and arms and anywhere else to feel cooler (or less 'ugh') for a short time afterwards. Use throughout the day or during long hot nights or fevers. Especially good in summer. (Avoid synthetic rose scented products. If smelling like a rose holds little appeal, try making your own very diluted ginger oil spray or other scented spray, or use plain water.)

- A bed bath can be given by a carer, or can be done by the patient using a bowl of warm, very mildly soapy water and a soft face washing cloth or sponge. It may not be possible to wash all areas in one attempt. Products such as pre-moistened body wipes may be used where the patients can tolerate the chemicals they contain. The advantage of these wipes is that no drying is needed afterwards. Some patients alternate the use of wipes with a soapy water and sponge bed bath.

- If you get up to go to the toilet, you could possibly do a 'drive-by sponge-bath' by washing a small part of yourself each time you go. One trip may include a quick face wash using a liquid soap, another toilet trip might include a quick underarm wash or tooth brushing.

- A warm soft and fluffy robe is a good idea for wearing on the trip from the bathroom to the bedroom, or for just after you have had a sponge bath or are changing clothes.

- Some patients recommend wiping under arms with hydrogen peroxide and then using a deodorant if you are unable to bathe every day.

- Talcum powder (if not allergic) or cornflour can be used on folds where skin cracks due to sweating and then doesn't heal. Cornflour also helps somewhat where skin is reacting to seams or parts of clothes.

- Changing clothes and sheets more often can help offset the problem of non-daily baths.

A note to M.E. patients: If you are not well enough to bathe daily at the moment, know that many severely and even moderately affected M.E. patients can bathe 'properly' only once a week or so, due to severe illness. This is very common in M.E., even though it isn't something that is often talked about. While it is not ideal, it is also nothing to be ashamed of.

Just do the best you can with your 'birdbaths' or sponge baths (and perhaps hair that sometimes looks like a birds *nest*!), hang in there and keep hoping for, and working toward, better health in the future, and know that you aren't alone.

More information

- To find more disability equipment that may be useful for making bathing easier, patients or carers may like to do an internet search for terms such as 'bathing disability equipment.'
- For more information on the severity of M.E. and caring for M.E. patients please see Hospital or carer notes for M.E. and Why patients with severe M.E. are housebound and bedbound.
- More tips on living with M.E. are available on the Practical tips for living with M.E. web page.

Acknowledgments
Thanks to Claire Bassett for editing this paper.

Thank you to (the late) Aylwin Catchpole, Ingeborg, Bea, Frances, Victoria, LK Woodruff, Clytie and everyone else that contributed to this paper. This paper was Aylwin's idea and she made several very helpful contributions to it, as she did to so many HFME papers. Sadly Aylwin (a long-term M.E. patient) died before this paper was completed.

Assisting the M.E. patient in managing toileting tasks

COPYRIGHT © JODI BASSETT OCTOBER 2010. UPDATED JUNE 2011. FROM WWW.HFME.ORG

 The problems with mobility and maintaining an upright posture previously described in 'Assisting the M.E. patient in managing bathing and hair care tasks' can also affect toileting in Myalgic Encephalomyelitis (M.E.) patients.

The physical exertion required to complete toileting tasks can at times be very difficult for moderately to severely affected patients with M.E. It may cause relapse or increased pain during and/or afterwards, and the tasks may have to be modified in some way to be completed or may require assistance from a carer.

M.E. patients may also face problems such as: loose stools or constipation (or both, at different times), urgency, frequency and urinary and/or faecal weakness or incontinence.

The following suggestions and tips come from people that have been ill with M.E. for a considerable time and would like to share some of the solutions they have found with their fellow patients, in the hopes that others may benefit from their years of experience.

Some tips are aimed at M.E. patients themselves, while others provide information appropriate for carers and parents of children with M.E.

Tips on devices that may assist the severely affected patient with toileting:

- If at all possible, the moderately to severely affected M.E. patients should have access to an ensuite bathroom, to save them wasting valuable time upright walking down a hallway etc. to the bathroom. This may also allow some patients to continue toileting themselves independently where this would otherwise not be possible.

- Some patients may find it helpful to have a bar attached to the wall of the toilet that can be used to help them lift their own body weight.

- Wheelchair using patients may need a wheelchair lift to be installed in order to help their carers place them on the toilet.

- Catheterisation carries a significant infection risk in M.E. patients and should be avoided unless there is no other choice.

- Some patients may require the help of a carer and the use of a bedpan for toileting.

- If patients are able to sit upright for a few minutes, a better choice than a bedpan may be a commode chair. This is a plastic or metal chair that can be used in the bedroom, and has a slide-out bedpan that can easily be removed for cleaning by a carer.

- Portable plastic containers designed for urinating in may be helpful for bedbound patients. Different designs are available for men and for women (such as the Urifem).
      Patients may be able to use and rinse out the containers themselves (or with assistance from a carer) and save their valuable time upright for trips to the toilet to defecate. Avoiding all the daily trips to the toilet for urination could mean the difference between improvement and deterioration for the M.E. patient. Using a regular toilet for defecation minimizes cleanup afterwards enormously and can also save the patients dignity.

- Raised toilet seats are available for those patients that find lifting themselves up from lower seats difficult.

- Patients that find regular toilet paper painful to use may benefit from having a bidet installed, or having a bidet attachment fitted to their regular toilet.

- Some severely affected patients may require the use of adult diapers, fitted and changed often with the help of a carer. Disposable underpads or waterproof sheets for the bed may also be helpful, along with absorbent pads for use inside a patient's underwear if incontinence is mild.

Additional notes:

- Kegel exercises may benefit many patients with urinary incontinence, as the exercises help to strengthen the pelvic floor muscles. However, Kegel exercises may be impossible for the M.E. patient to complete or even counter-productive due to muscle weakness and paralysis that occurs in M.E. when any muscle is used repeatedly. These exercises may in fact make the problem much worse unless the patient is well enough to tolerate mild exercise.

- If the patient is very ill and basic tasks such as walking to the toilet half a dozen times a day are almost all that can be done each day, and only then with great difficulty, it may well be worth considering the use of a Urifem or a portable male urinal. Giving up independent and unmodified toileting while still able is very difficult emotionally, but may really pay off in the long term. If the patient is overexerting each day then disease progression and deterioration becomes more and more a reality and so pre-empting a forced change of toileting habits may in fact be the best way to ensure independent toileting in the long term.

- It should be noted that patients very severely affected with M.E. may be in the terrible position of being too ill to tolerate any other form of toileting than constant diaper use, but also too ill to tolerate – without extreme pain and suffering from the noise, movement and the interaction with another person – the carer regularly changing and refitting the diapers. Some patients are also so ill that although they need this type of constant care, they do not qualify for it.

  M.E. patients in either or both of these positions are living in a situation more terrible than most people can even imagine and are a prime reason why some genuine advocacy is so desperately needed, right now. Please see What is M.E.? for more information.

More information

- To find more disability equipment that may be useful for making toileting easier, patients or carers may like to do an internet search for terms such as 'toileting disability equipment' or 'urifem' or 'bidet attachment.'
- For more information on the severity of M.E. and caring for M.E. patients please see Hospital or carer notes for M.E. and Why patients with severe M.E. are housebound and bedbound.

Acknowledgments
Thanks to Claire Bassett for editing this paper. Thank you to everyone that contributed to this paper.

Permission is given for each of the individual papers in this book to be freely redistributed by email or in print for any genuine not-for-profit purpose provided that the entire text, including this notice and the author's attribution, is reproduced in full and without alteration.

The HFME 3 part M.E. ability and severity scale

COPYRIGHT © JODI BASSETT JUNE 2005. UPDATED JUNE 2011. FROM WWW.HFME.ORG

3 PART M.E. ABILITY & SEVERITY SCALE: PART 1 – PHYSICAL ABILITY SCALE

FULLY RECOVERED

100%   A pre-illness level of physical activity is possible.

VIRTUALLY RECOVERED

90%   A high level of physical capabilities (around 90%): full-time study or work without difficulty is achievable in addition to a full and active social life.

MILDLY AFFECTED

80%   A high level of physical activity is possible (around 80%) with minimal restrictions involving exertion. Patient is capable of working full time in jobs not requiring exertion.

70%   Physical activity is at/or around 70%. A daily activity limit is clearly noted. Incapable of full-time work in jobs requiring physical exertion, but able to work full-time in lighter activities if hours are flexible. Social life is restricted to non-exertive activities.

MODERATELY AFFECTED

60%   Physical activity is at/or around 60%: strenuous activities are difficult, but light activities and desk work are achievable as long as the total time worked is 5 – 7 hours a day and regular rest periods are observed. Physical abilities degenerate significantly with sustained exertion.

50%   Physical activity is at around 50%: part-time work, light activities or desk work are acceptable for up to 4 - 5 hours a day as long as requirements for quiet and rest are met. Physically undemanding social activities are possible. Physical abilities degenerate significantly with sustained exertion. Unable to perform strenuous tasks.

MODERATELY TO SEVERELY AFFECTED

30%   Overall activity level reduced to at/or around 30 - 40%. May be unable to walk without support much beyond 100/200 metres; a walking stick or wheelchair may be used to travel longer distances. Several hours of desk work may be possible each day if requirements for quiet and resting are met.

Physically undemanding social activities are possible.

20%   Overall physical activity level reduced to around 20%. Not confined to the house but may be unable to walk without support much beyond 50/100 metres; a wheelchair may be used to travel longer distances. Requires 3 or 4 regular rest periods during the day; only one 'large' activity possible per day usually requiring a day or more of rest.

(A large activity is individual; it could be cleaning cupboards or having visitors; it is any activity that the patient finds difficult and so no longer considers 'usual.')

SEVERELY AFFECTED

10%   Overall physical activity level reduced to around 10%. Confined to the house but may occasionally (and with a significant recovery period) be able to take a short wheelchair ride or walk, or be taken to see a doctor. Most of the day needs to be spent resting except for a period of several hours interspersed throughout the day when small tasks may be completed (or one larger one). Activity is mostly restricted to managing the tasks of daily living where some assistance is needed and modification of tasks may be required.

5%   Overall physical activity level reduced to around 5%. Usually confined to the house but may very occasionally (with a recovery period of a week or more) be able to take a short wheelchair ride or walk, or be taken to see a doctor. Bed-bound or couch-bound for 21+ hours a day. Activity is restricted almost exclusively to managing the tasks of daily living where some assistance with modification of tasks is necessary.

VERY SEVERELY AFFECTED

3%   Overall physical activity level severely reduced to around 3%. No travel outside the house is possible. Bed-bound the majority of the day (22+ hours) but may (with difficulty and an exacerbation of symptoms) be able to sit up, walk or be pushed in a wheelchair for very short trips within the home. Nearly all tasks of daily living need to be performed and/or heavily modified by others. Due to problems with swallowing, eating may be very difficult.

1%   Overall physical activity level very severely reduced to around 1%. No travel outside the house is possible. Close to completely bed-bound (lying flat in bed 23.5+ hours a day). May sometimes (with difficulty and with an exacerbation of symptoms) be able to sit up, walk or be pushed in a wheelchair within the home. All tasks of daily living need to be performed and/or very heavily modified by others. Eating and drinking may be very difficult.

PROFOUNDLY SEVERELY AFFECTED

0.5%   Completely bed-bound and may be unable to turn or move at all. Eating is extremely difficult and liquid food may be necessary (little and often). When swallowing becomes difficult, nasal feeding tubes may be required. Unable to care for ones self at all; bed baths and other personal care that are undertaken by a care-giver may cause a severe relapse in symptoms and/or disease progression and so should not automatically be attempted every day.

3 PART M.E. ABILITY & SEVERITY SCALE: PART 2 – COGNITIVE ABILITY SCALE

FULLY RECOVERED

100%   An unrestricted level of cognitive functioning is possible.

VIRTUALLY RECOVERED

90%   A high level of cognitive functioning is possible; around 90% of pre-illness level. Able to cope on a cognitive level with full-time study or work without difficulty and enjoy a full social life.

MILDLY AFFECTED

80%   A high level of cognitive functioning is possible, around 80 - 90%. Minimal restrictions apply for activities that demand a high standard of cognitive functioning. Unable to manage full-time study or work without difficulty in areas that place an excessive demand on a cognitive level.

70%   Cognitive functioning is at/or around 70 - 80%; a daily cognitive activity limit is clearly noted. Unable to work fulltime where high demands are made on a cognitive level, but can work fulltime in less demanding jobs if hours are flexible. Some restrictions on social life.

MODERATELY AFFECTED

60%   Cognitive functioning is at/or around 60% ; unable to perform tasks which are excessively demanding on a cognitive level, but can complete lighter activities for 5 – 7 hours a day although rest periods are required.

Cognitive functioning degenerates significantly in a crowded, noisy or busy environment or with sustained and/or high level use. Social life may be moderately affected.

50%   Cognitive functioning is at/or around 40 -50%; unable to perform tasks which are excessively demanding on a cognitive level, but able to work part-time in lighter activities for 4 - 5 hours a day (or perhaps longer at a reduced quality level) if requirements for quiet and resting are met.

Cognitive functioning degenerates significantly in a crowded, noisy or busy environment or with sustained and/or high level use. Social activities with environments that are quiet and not mentally challenging are possible.

MODERATELY TO SEVERELY AFFECTED

30%   Cognitive functioning is reduced to around 30 - 40%; unable to perform mentally challenging tasks, but able to complete simpler cognitive tasks (study or work) for 3 – 4 hours a day (or perhaps longer at a lower quality level) if requirements for quiet and resting are met.

Concentration and cognitive ability are significantly affected. Following the plots of some TV shows or books may be difficult. Non-mentally challenging social activities are possible on a limited basis.

20%   Cognitive functioning is reduced to around 20%; unable to perform mentally challenging tasks easily or often, but able to complete less complex cognitive tasks for 2 – 3 hours a day (or perhaps longer at a lower quality level) if requirements for quiet and resting are met.

Concentration, memory and other cognitive abilities are significantly affected. Following the plots of TV shows or books may be difficult.

Non-mentally challenging social activities possible on a limited basis.

SEVERELY AFFECTED

10%   Cognitive functioning is reduced to around 10%; unable to perform mentally challenging tasks easily or often, but able to complete less complex cognitive tasks for 1 – 2 hours a day (or perhaps longer at a lower quality level) if requirements for quiet and resting are met.

Concentration, memory and other cognitive abilities are significantly affected at all times and may be severely affected during relapses. Concentration for more than half an hour at a time may be extremely difficult. Following the plots of some TV shows or books may be difficult or impossible. Non-mentally challenging social activities possible on a very restricted basis.

5%   Cognitive functioning is reduced to around 5%; unable to perform even moderately mentally challenging tasks easily or often, but able to complete less complex cognitive tasks for about an hour or so each day (or perhaps longer at a lower quality level) if requirements for quiet and resting are met.

Concentration, memory and other cognitive abilities are significantly affected at all times and may be severely affected during relapses. Concentration for more than 10 to 15 minutes at a time may be extremely difficult. Following the plots of TV shows or books may be difficult or impossible. Non-mentally challenging social activities possible occasionally for short periods.

VERY SEVERELY AFFECTED

3%   Cognitive functioning is reduced to less than 5%; able to complete simple cognitive tasks for about 10-30 minutes each day (or perhaps longer at a lower quality level) if requirements for quiet and resting are met.

Concentration, memory and other cognitive abilities are severely affected. Concentration may be extremely difficult. Only short periods of TV, radio or reading are possible. A friend can be seen for approximately 10 - 30 minutes a week.

1%   May be able to complete simple cognitive tasks such as talking, listening to speech or reading (with difficulty) for several 2–10 minute periods throughout the day if requirements for quiet and resting are met.

Concentration, memory and other cognitive abilities are very severely affected. Concentration may be extremely difficult. There may be an inability to maintain full consciousness throughout the day. No TV is possible but quiet music or an audio book may be listened to for short periods. A close friend or family member can be seen for a few minutes, occasionally.

PROFOUNDLY SEVERELY AFFECTED

0.5%   Concentration, memory and other cognitive abilities are extremely and severely affected. Achieving even a low level of concentration may be extremely difficult or impossible, and there may be a high degree of cognitive confusion as a result. No TV or radio is possible. There may also be a difficulty maintaining consciousness for more than a few minutes at a time. Receiving visitors (even close family members) is almost impossible or impossible. Talking, reading or writing more than the occasional few words is often impossible.

FULLY RECOVERED

100%   No symptoms

VIRTUALLY RECOVERED

90%   No symptoms at rest. Mild symptoms on occasion following strenuous physical or mental activity but recovery is complete by the next day.

MILDLY AFFECTED

Note that symptom severity on a scale of one to ten means:
Mild Symptoms = 1 to 3. Symptoms present but at so low a level you can forget they are there most of the time.
Mild/moderate symptoms = 4 to 5
Moderate symptoms = 6 to 7
Very Severe Symptoms = 8
Severe Symptoms = 9
Extremely Severe Symptoms = 10. Totally non-functional. Absolute agony.

80%   No symptoms at rest. Mild symptoms (1 to 3) for several hours or days following strenuous physical or mental activity.

70%   Mild symptoms (1 to 3) at rest, worsened to mild/moderate (4 or 5) for several hours or days following strenuous physical or mental activity beyond the person's limits.

MODERATELY AFFECTED

60%   Mild - mild/moderate symptoms (1 to 5) at rest, worsened to moderate (6 or 7) for several hours or days following physical or mental activity beyond the person's limits.

50%   Mild/moderate symptoms (4 or 5) at rest, consisting of mild/moderate pain and/or sensations of illness/dysfunction throughout the body and brain for some parts of the day, with increasing moderate symptoms (6 or 7) for several hours, days or weeks (or longer) following physical or mental activity beyond the person's limits.

MODERATELY TO SEVERELY AFFECTED

30%   Moderate symptoms (6 or 7) at rest with moderate pain and/or sensations of illness/dysfunction throughout the body and brain for significant periods of the day; increasing moderate (and occasionally severe – level 8) symptoms for several hours, days or weeks or months (or longer) following physical or mental activity beyond the persons limits.

20%   Moderate (6 or 7) and occasionally severe (8) symptoms at rest. There is moderate pain (6 or 7) and/or sensations of illness/dysfunction throughout the body and brain for significant periods of the day, increasing to moderate and sometimes severe symptoms for several hours, days, weeks or months (or longer) afterward.

SEVERELY AFFECTED

10%   Moderate to severe symptoms (6–8) at rest. There is moderate to severe pain (6–8) and/or sensations of illness/dysfunction throughout the body and brain for much of the day. Symptoms are severe (8) following any physical or mental activity with a recovery period as low as hours, or as long as days to months, or longer. It is all the person can do to just get through one day at a time.

5%   Severe symptoms (8) at rest and following even trivial physical or mental activity with a recovery period of hours or days, or as much as several weeks or months or longer. There is severe pain (8) and/or overwhelming sensations of illness/dysfunction throughout the body and brain for all but a few hours of the day. In some patients only small amounts of stimuli can be tolerated, and only for short periods of time. It is all the person can do to just get through the day a few hours at a time.

VERY SEVERELY AFFECTED

3%   There is severe pain (8) and/or overwhelming sensations of illness/dysfunction throughout the body and brain for all but a few short periods in the day, increasing to severe or very severe symptoms (8 or 9) following even trivial physical or mental activity with a recovery period of hours, days, weeks, months or longer. In some patients only small amounts of stimuli can be tolerated for short periods. It is all the person can do to just get through the day one hour at a time.

1%   There is severe pain (8) and/or overwhelming sensations of illness/dysfunction throughout the body and brain almost continually, worsening to very severe (9) or extremely severe (10) following even trivial physical or mental activity with a recovery period of hours, days, weeks, months or longer. In some patients any type of stimulus is intolerable; even very low levels of light, noise, movement or motion are excruciating for more than very short periods. The smallest physical movements bring extreme exacerbation of symptoms. Intellectual activity is similarly affected. It is all the person can do to just get through the day one minute at a time.

PROFOUNDLY SEVERELY AFFECTED

0.5%   There is very severe (9) pain and/or overwhelming sensations of illness/dysfunction throughout the body and brain *continually,* worsening to extremely severe (10) by even trivial physical or mental activity with a recovery period of hours, days, several weeks, months or longer. In some patients any type of stimulus is intolerable; even very short/low exposures to light, noise, movement and motion are excruciating and may require a long recovery period. The smallest physical movement brings intense exacerbation of symptoms. Mental activity is similarly affected. It is all the person can do to just get through the day one minute or one second at a time.

These scales are not intended for medical use and were not created by a medical professional. They are designed to be used by Myalgic Encephalomyelitis (M.E.) sufferers, and perhaps also their carers, to measure improvements and changes in different aspects of the illness over time.

Because physical and cognitive ability and symptom severity are often not equally affected in every patient, this scale is divided into three parts.

A scale with more than one category should ensure greater accuracy and hopefully be more encouraging since there is a great likelihood that patients will score a bit higher in at least one category compared to the other two.

Terminology used in the scales
Resting

Resting means completely different things at different severity levels of illness. For the *mildly ill*, resting may be watching TV or sitting in a chair while reading a book or having a quiet visit with friends. For the *severely ill*, these activities are not at all restful and indeed would provoke relapses.

For the *very severely ill*, resting means lying down in a dark room in silence and with no sensory input at all (TV, radio or light) with zero physical movement or cognitive activity. Clothing must also be comfortable and the room must have a very moderate temperature; not too hot or cold. When referring to resting, a better term for *the very severely ill* would be '*complete incapacitation.*'' The term 'resting' implies that inactivity is optional; the *severely ill* are often 'resting' (i.e. incapacitated) because it is physically impossible for them to do anything else.

For *moderately ill* patients, resting lies somewhere between the above two extremes.

Resting will change according to the severity level of each individual. The *very severely ill* have no symptom–free or safe activity limit. Concepts of pacing or of keeping activity at a level which does not cause immediate or delayed symptoms are useless. Indeed, a sizeable proportion of the *very severely ill* may well be so affected in the first place **because of over-exertion in the early stages of their illness;** they did not know the importance of rest, and in some cases, were not *allowed* to rest adequately. Extremely common in M.E., this is a tragedy and an absolute disgrace.

Severe M.E. restricts life to a degree that healthy people might find hard to imagine, but patients have learned from bitter experience all about the negative consequences of over-exertion. They are reminded on a weekly or daily basis that even with careful control, limits can be misjudged or tasks can take a greater toll than expected. An M.E. patient can never be accused of being too restrictive of her activity levels; she wants to live and experience life as much as possible and has learned to use enormous discipline to avoid over-exertion.

I have never heard of anyone with M.E. who is *too* restrictive with their activity levels; the problem is always the opposite. It is a natural human desire to "keep going" when there are chores waiting to be completed, and for most patients, when there is a moment of feeling somewhat more 'well,' it is often less difficult to physically keep pushing themselves (even to the point of worsening the illness) than it is to force themselves to adequately rest. In the earlier stages of the disease (when pushing oneself for short periods is more possible) the patient also often finds it is easier emotionally to forge ahead in physical over-exertion and suffer the consequences rather than stand up to extreme pressure from friends, family and medical staff for these activities to be completed at the same level as in pre-illness..

Resting endlessly for many years is much harder than one can imagine. It has been observed that it is less difficult for a stroke victim to learn to walk and talk again than it is for an M.E. patient to discipline herself to rest endlessly, with no distraction from the chronic pain.

People with M.E. would give anything to be able to work hard to improve their illness, and to be improving every day instead of staying the same or getting worse. The problem of M.E. patients *under-reporting* or *under-estimating* their ability levels just *does not exist*.

This is not about patients being as inactive as possible. A person with moderate M.E. of course does not need to live with the same restrictions as does someone with severe M.E. The point here is just that patients must stay within their individual post-illness limits.

Many M.E. patients try to adhere to the 80% rule. The idea of the 80% rule is for patients to work out how much they can do every day without becoming in any way sicker, and then do only 80% of that. To have each day be the same activity-wise is the goal – without cycles of adrenaline surges and relapses. Avoiding overexertion is essential, but it is not enough. Getting some real rest is important for the M.E. patient too, so that the body has some extra energy and resources to use for healing.

M.E. patients that aren't sure if they are resting enough may try resting significantly more for a week to see which symptoms improve, if any. If symptoms improve then the patient likely needs to cut back activities. Increasing the activity levels of someone with M.E. beyond their individual limits can only be harmful. Gradually increasing activity does not make exercise easier or less damaging; it does not matter how slowly it is done: it has the same harmful effects.

Overexertion

What characterises M.E. every bit as much as the individual symptoms is the way in which people with M.E. respond to physical and cognitive activity, sensory input and orthostatic stress.

The main characteristics of the pattern of symptom exacerbations, relapses and disease progression in M.E. include the following:

A. People with M.E. are unable to maintain their pre-illness activity levels. This is an acute (sudden) change. M.E. patients can only achieve 50% or less of their pre-illness activity levels.

B. People with M.E. are limited in how physically active they can be but are also limited in similar ways with cognitive exertion, sensory input and orthostatic stress.

C. When a person with M.E. is active beyond their individual physical, cognitive, sensory or orthostatic limits, this causes a worsening of various neurological, cognitive, cardiac, cardiovascular, immunological, endocrinological, respiratory, hormonal, muscular, gastrointestinal and other symptoms.

D. The level of physical activity, cognitive exertion, sensory input or orthostatic stress that is needed to cause a significant or severe worsening of symptoms varies from patient to patient, but is often trivial compared to a patient's pre-illness tolerances and abilities.

E. The severity of M.E. waxes and wanes throughout the hour/day/week and month.

F. The worsening of the illness caused by overexertion often does not peak until 24 - 72 hours (or more) later.

G. The effects of overexertion can accumulate over longer periods of time and lead to disease progression or death.

H. The activity limits of M.E. are not short term; an increase in activity levels beyond a patient's individual limits, even if gradual, causes relapse, disease progression or death.

I. The symptoms of M.E. do not resolve with rest. The symptoms and disability of M.E. are not caused only by overexertion; there is also a base level of illness which can be quite severe even at rest.

J. Repeated overexertion can harm the patient's chances for future improvement in M.E. Patients who are able to avoid overexertion have repeatedly been shown to have the most positive long-term prognosis.

K. Not every M.E. sufferer has 'safe' activity limits within which they will not exacerbate their illness; this is not the case for very severely affected patients.

Sensory input

Sensory input includes light, noise, movement, motion, vibration, odour and touch.

Cognitive abilities

When rating cognitive abilities it is the person's intellectual capabilities which are being referred to, not their state of mental or emotional health which will most often be at quite a different level altogether.

Symptom severity

Mild Symptoms = 1 to 3. Symptoms present but at so low a level you can forget they are there most of the time.
Mild/moderate symptoms = 4 to 5
Moderate symptoms = 6 to 7
Very Severe Symptoms = 8
Severe Symptoms = 9
Extremely Severe Symptoms = 10. Totally non-functional and/or being near delirium. Completely engulfed in, and overwhelmed with, pain. Absolute agony.

The pain and suffering of M.E. have a number of different 'flavours.' The experience can be made up of severe nausea, vertigo and disequilibrium, cold and hot fevers or feeling both very cold and very hot at the same time, feeling 'poisoned' and very ill, pain in the glands and throat, muscle pain, twitching and uncontrollable spasms, difficulty breathing and breathlessness, cardiac pain and pressure and dysfunction that feels like a heart attack, a feeling of having a heart attack in every organ (caused by lack of blood flow to these organs), sensations of pain and terrible pressure in the brain and behind the eyes, stroke-like or coma-like episodes, abdominal pain and pain/discomfort following meals, seizures and 'sensory storms' (while conscious) and, lastly, an inability to remain conscious for more than a few minutes or hours at a time or for more than a few hours each day in total. Any one of these problems can cause extreme suffering. What makes severe M.E. so terrible is that the patient is almost always dealing with a large number of these problems *all at once*.

For more (fully referenced) information on M.E., see: What is M.E.? plus Why patients with severe M.E. are housebound and bedbound and The importance of avoiding overexertion in M.E. See also: M.E. vs. M.S.: Similarities and differences

Suggestions on using these scales:

Patients and carers can make charting progress as simple or as complicated as is desired:

- Simple charting: Every few months, write down your (or the patient's) scores on each of the three scales along with the date.
- Detailed charting: Have a chart that includes "good day" and "bad day" columns, and a notations section, and fill them in accordingly, as shown in the example below.

DATE:	GOOD DAY	BAD DAY	NOTATIONS:
10/10/2011	40% overall		To have days this good I need to rest almost totally for weeks beforehand including avoiding all trips out of the house and any other serious exertions.
10/10/2011		20% 20% 10%	I scored 20% on the cognitive ability scale but my ability to handle sensory input is about 10%.

Few people will find that this or any other chart describes their exact combination of symptoms or experience of the illness, so patients might find that modifying the digital download of the chart enables them to more fully describe their own symptoms.

To read a fully-referenced version of the medical information in this text compiled using information from the world's leading M.E. experts, please see the 'What is M.E.?' paper on page 113 of this book or on the HFME website.

Acknowledgments
Thanks to Roseanne Schoof for editing this paper.

Relevant quotes
'[A] crucial differentiation between M.E. and post viral fatigue syndrome lies in the striking variability of the symptoms not only in the course of a day but often within the hour. This variability of the intensity of the symptoms is not found in post viral fatigue states.'
DR MELVIN RAMSAY

The HFME ability scale for M.E. patients: Summary

COPYRIGHT © JODI BASSETT 2005. UPDATED JUNE 2011. FROM WWW.HFME.ORG

A form to be filled out by the M.E. patient for the benefit of hospital staff and carers

My name is _____

I have had Myalgic Encephalomyelitis (M.E.) for _____
months/years, since I was _____ years old.

When I was first ill, I was (circle the correct answer) mildly/moderately affected, moderately affected, moderately/severely affected, severely affected, very severely affected.

I am currently (tick the appropriate box):

☐ MODERATELY AFFECTED

Physical activity is at around 50%: part-time work, light activities or desk work are acceptable for up to 4 - 5 hours a day as long as requirements for quiet and rest are met. Physically undemanding social activities are possible. Physical abilities degenerate significantly with sustained exertion. Unable to perform strenuous tasks.

Cognitive functioning is at/or around 40 -50%; unable to perform tasks which are excessively demanding on a cognitive level, but able to work part-time in lighter activities for 4 - 5 hours a day (or perhaps longer at a reduced quality level) if requirements for quiet and resting are met. Cognitive functioning degenerates significantly in a crowded, noisy or busy environment or with sustained and/or high level use. Social activities with environments that are quiet and not mentally challenging are possible.

Mild/moderate symptoms (4 or 5 out of 10) at rest, consisting of mild/moderate pain and/or sensations of illness/dysfunction throughout the body and brain for some parts of the day, with increasing moderate symptoms (6 or 7) for several hours, days or weeks (or longer) following physical or mental activity beyond the person's limits.

☐ MODERATE TO SEVERELY AFFECTED (1)

Overall activity level reduced to at/or around 30 - 40%. May be unable to walk without support much beyond 100/200 metres; a walking stick or wheelchair may be used to travel longer distances. Several hours of desk work may be possible each day if requirements for quiet and resting are met. Physically undemanding social activities are possible.

☐ MODERATE TO SEVERELY AFFECTED (2)

Overall physical activity level reduced to around 20%. Not confined to the house but may be unable to walk without support much beyond 50/100 metres; a wheelchair may be used to travel longer distances. Requires 3 or 4 regular rest periods during the day; only one 'large' activity possible per day usually requiring a day or more of rest. (A large activity is individual; it could be cleaning cupboards or having visitors; it is any activity that the patient no longer considers 'usual.')

☐ SEVERELY AFFECTED

Overall physical activity level reduced to around 10%. Confined to the house but may occasionally (and with a significant recovery period) be able to take a short wheelchair ride or walk, or be taken to see a doctor. Most of the day needs to be spent resting except for a period of several hours interspersed throughout the day when small tasks may be completed (or one larger one). Activity is mostly restricted to managing the tasks of daily living where some assistance is needed and modification of tasks may be required.

☐ VERY SEVERELY AFFECTED

Overall physical activity level severely reduced to around 3%. No travel outside the house is possible. Bed-bound the majority of the day (22+ hours) but may (with difficulty and an exacerbation of symptoms) be able to sit up, walk or be pushed in a wheelchair for very short trips within the home. Nearly all tasks of daily living need to be performed and/or heavily modified by others. Due to problems with swallowing, eating may be very difficult.

☐ EXTREMELY SEVERELY AFFECTED

Overall physical activity level very severely reduced. No travel outside the house is possible. Close to completely bedbound (lying flat in bed 23.5+ hours a day). May sometimes (with difficulty and with an exacerbation of symptoms) be able to sit up, walk or be pushed in a wheelchair for very short periods/distances interspersed throughout the day (to the bathroom or to travel from room to room). All tasks of daily living need to be done by others and/or very heavily modified. Eating and drinking may be very difficult.

☐ PROFOUNDLY SEVERELY AFFECTED

Completely bed-bound and may be unable to turn or move at all. Eating is extremely difficult and liquid food may be necessary (little and often). When swallowing becomes difficult, nasal feeding tubes may be required. Unable to care for ones self at all; bed baths and other personal care that are undertaken by a care-giver may cause a severe relapse in symptoms and/or disease progression and so should not automatically be attempted every day.

Concentration, memory and other cognitive abilities are severely affected. Achieving even a low level of concentration may be extremely difficult or impossible, and there may be a high degree of cognitive confusion as a result. No TV or radio is possible. There may also be a difficulty maintaining consciousness for more than a few minutes at a time. Receiving visitors (even close family members) is almost impossible or impossible. Talking, reading or writing more than the occasional few words is often impossible.

There is very severe (9) pain and/or overwhelming sensations of illness/dysfunction throughout the body and brain *continually,* worsening to extremely severe (10) by even trivial physical or mental activity with a recovery period of hours, days, months or longer. In some patients any type of stimulus is intolerable; even very short/low exposures to light, noise, movement and motion are excruciating and may require a long recovery period. The smallest physical movement brings intense exacerbations in symptoms. Mental activity is similarly affected. It is all the person can do to just get through the day one minute or one second at a time.

Please see the full-length version of this text for more information.

Additional patient notes (if necessary):

The HFME M.E. ability and severity scale checklist

COPYRIGHT © JODI BASSETT AUGUST 2010. FROM WWW.HFME.ORG

M.E. patients may circle the correct number or fill in the blank spaces below.

Name: _____ Date: _____

This checklist refers to my abilities on (1) an average day/week, (2) my worst day/week or (3) my best day/week.

Average time spent upright (standing or sitting) daily: 1 2 3 4 5 6 7 8
KEY: (1) 0 – 5 mins (2) 5 -15 mins (3) 15 - 30 mins (4) 30 - 60 mins (5) 1 - 2 hrs (6) 2 - 3 hrs (7) 4 - 5 hrs (8) 6 - 7 hrs (9) 8 - 10 hrs (10) 12 + hrs

Average time spent on the computer daily: 1 2 3 4 5 6 7 8 9 10

Average time spent with another person or talking on the phone daily: 1 2 3 4 5 6 7 8 9 10

Average time spent reading daily: 1 2 3 4 5 6 7 8 9 10

Average time spent listening to music or audio books daily: 1 2 3 4 5 6 7 8 9 10

Average time spent watching TV daily: 1 2 3 4 5 6 7 8 9 10

Average amount of time high-quality level thinking is possible daily: 1 2 3 4 5 6 7 8 9 10

Average time spent doing housework daily: 1 2 3 4 5 6 7 8 9 10

Average time spent in a total rest state (lying in a darkened quiet room) daily: 1 2 3 4 5 6 7 8 9 10

Ability to eat and drink (ability to use utensils, chew and swallow): (1) I need to be fed entirely by tube (2) very poor (3) poor (4) average (5) good (6) almost always excellent (7) excellent, no problems at all.

Frequency of trips out of the house: (1) housebound for years now (2) housebound for many months now (3) rare (4) once every few months (5) once every 2 – 4 weeks (6) once a week (7) twice weekly (8) 3 – 4 times weekly.

Ability to bathe unassisted: 1 2 3 4 5 6 7 8
KEY: (1) a sponge or bed bath is possible less than once a week (2) a sponge or bed bath is possible once a week (3) a shower or bath is possible once a week so long as the task is modified in certain ways (4) a sponge or bed bath is possible 3 – 5 times weekly (5) a daily sponge or bed bath is possible (6) a bath or shower is possible 3 – 5 times weekly so long as the task is modified in certain ways (7) a daily bath or shower is possible so long as the task is modified in certain ways (8) a daily bath or shower is possible with no problems or restrictions.

Ability to bathe with assistance from a carer: 1 2 3 4 5 6 7 8

Average waking time: _____ am/pm Average time sleep is initiated: _____ am/pm.

Average time spent napping in the daytime: _____

Sleep quality and duration is: (1) appalling (2) very poor (3) poor (4) average (5) good (6) very good (7) excellent.

Sensitivity to noise, rated from 1 to 10, with 10 being extremely severe: 1 2 3 4 5 6 7 8 9 10

Sensitivity to light, rated from 1 to 10: 1 2 3 4 5 6 7 8 9 10

Vertigo and balance problems, rated from 1 to 10: 1 2 3 4 5 6 7 8 9 10

Pain, discomfort and physical suffering, rated from 1 to 10: 1 2 3 4 5 6 7 8 9 10

Nausea, rated from 1 to 10: 1 2 3 4 5 6 7 8 9 10

Sensitivity to (viewing) movement, rated from 1 to 10: 1 2 3 4 5 6 7 8 9 10

Burning eye pain/blurred vision, rated from 1 to 10: 1 2 3 4 5 6 7 8 9 10

General neurological problems and symptoms, rated from 1 to 10: 1 2 3 4 5 6 7 8 9 10

Severity of cardiac episodes (pain/abnormal function/very low blood pressure), rated from 1 to 10: 1 2 3 4 5 6 7 8 9 10

Frequency of severe cardiac episodes, rated from 1 to 10: 1 2 3 4 5 6 7 8 9 10

On average I overexert to the point of a mild-moderate worsening of symptoms and/or adrenaline burst (a hyperactive state caused by the body being put in significant physiological difficulty): (1) every day (2) almost every day (3) every few days (4) once a week (5) once a fortnight (6) every 2 – 3 weeks (7) every 4 – 6 weeks (8) rarely (9) almost never (10) I don't do this at all anymore.

On average I overexert to the point of a moderate-severe worsening of symptoms and/or adrenaline burst (a physiological difficulty caused hyperactive state): 1 2 3 4 5 6 7 8 9 10

The main *reasons* I overexert myself, if I do, are (*circle all that apply*): (1) I have no choice, I am not getting any of the care I need to stop my condition deteriorating (2) I need significantly more care than I am getting (3) I need a little bit more care than I am getting (4) I misjudge how able I am to do certain tasks sometimes (5) I need to do some fun things sometimes even though I am too ill for them (6) arguments and heated discussions with some family members (7) dealing with and talking to my carers (8) talking to my friends and family members (9) caring for my children/partner (10) caring for my pets/garden (11) sorting out my finances and other paperwork/responsibilities (12) getting and giving support to my fellow patients (13) I get carried away doing fun tasks sometimes and choose to keep going longer than I should despite the consequences (14) seeking medical care or trying to secure basic welfare (or insurance) entitlements.

My best time of day is _____

My worst is _____

Overall I feel my condition is: (1) at risk of becoming fatal (2) profoundly severe (3) very severe (4) severe (5) moderate/severe (6) moderate (7) mild/moderate (8) mild (9) in an almost complete remission (10) in remission.

Overall I feel my condition is (1) worsening rapidly and terrifyingly (2) worsening slowly (3) stable (4) stable in some ways and improving slowly in some ways (5) uncertain: some parts are worse, some the same and some better (6) improving very slowly in very small ways (7) improving significantly over time (8) improving at an accelerated rate. This is occurring primarily because (*leave blank if cause is unknown*):

_____

_____

_____

Additional notes (if necessary):

Suggested ways to use the HFME M.E. ability and severity scale checklist

- Patients are advised to print out lots of copies of the checklist, excluding this section, all at once and to fill one out once every 3 - 4 months, or whenever something changes worth making note of. Copies of the checklist can be downloaded free from the HFME website. Any questions that don't apply, just leave blank, or delete from the digital file. Patients may also choose to use only the first page of the checklist.

- When filling in the form, just circle the answers that are most correct and fill in the spaces appropriately OR do this and also write some notes in the margins giving additional information about some or all of the questions.

- This checklist can be used together with the more general HFME 3-part M.E. ability and severity scale, as they each serve slightly different functions.

- Note that the first key included in the form is designed to be used for a number of different questions (to save space), so not all numbered responses will be appropriate for all questions. (It is extremely unlikely that anyone with M.E. would spend 12 hours a day cleaning or on the computer, for example.)

Acknowledgments
Thanks to Claire Bassett for editing this paper.

Permission is given for each of the individual papers in this book to be freely redistributed by email or in print for any genuine not-for-profit purpose provided that the entire text, including this notice and the author's attribution, is reproduced in full and without alteration.

CHAPTER FOUR
More information about M.E.

This chapter includes the following two papers:

1. M.E. vs. M.S.: Similarities and differences – Condensed/modified version

Myalgic Encephalomyelitis (M.E.) and Multiple Sclerosis (M.S.) are very similar diseases medically in many ways. However, for reasons that have nothing to do with science, the two diseases are treated very differently politically and socially. The contrast could not be more stark.

M.E. patients are treated terribly (and often abused, even unto death in some cases), yet there is no public outcry as there would be if M.S. patients were treated in this same way. Thus people with M.E. find themselves in the position of actually envying people who have M.S.

2. What is M.E.? The full-length version of the paper.

A fully referenced historical, political and medical overview of M.E.

M.E. vs. M.S.: Similarities and differences - Condensed/modified version

 As many members of the public and the medical profession will be aware, Multiple Sclerosis (M.S.) is a disabling neurological disease which also affects the muscles. M.S. is a terrible disease and can cause severe disability and extreme suffering.

However, as surprising or bizarre as it seems, there is a section of the community which has reason to be envious of people who have M.S. It is made up of people who have the disabling neurological disease called Myalgic Encephalomyelitis (M.E.)

Medical similarities between M.S. and M.E.

M.E. and M.S. are actually very similar medically in many ways, as the following list demonstrates.

Table 1. Medical similarities between M.S. and M.E.	
Multiple Sclerosis	**Myalgic Encephalomyelitis**
M.S. is primarily a neurological disease, i.e. a disease of the central nervous system (CNS).	M.E. is primarily a neurological disease, i.e. a disease of the central nervous system (CNS).
Demyelination (damage to the myelin sheath surrounding nerves) has been documented in M.S.	Demyelination (damage to the myelin sheath surrounding nerves) has been documented in M.E.
Evidence of oligoclonal bands in the cerebrospinal fluid has been documented in M.S.	Evidence of oligoclonal bands in the cerebrospinal fluid has been documented in M.E.
No single definitive laboratory test is yet available for M.S. but a series of tests are available which can objectively confirm the diagnosis with some certainty.	No single definitive laboratory test is yet available for M.E. but a series of tests are available which can objectively confirm the diagnosis with a high degree of certainty.
M.S. can be severely disabling and cause significant numbers of patients to be bedbound or wheelchair-reliant.	M.E. can be severely disabling and cause significant numbers of patients to be bedbound, wheelchair-reliant or housebound.
M.S. can be fatal (either from the disease itself or from complications arising from the disease).	M.E. can be fatal (either from the disease itself or from complications arising from the disease).
M.S. significantly reduces life expectancy.	M.E. significantly reduces life expectancy.
Symptoms/problems which occur in M.S. include: impaired vision, nystagmus, afferent pupillary defect, loss of balance and muscle coordination,	Symptoms/problems which occur in M.E. include: impaired vision, nystagmus, afferent pupillary defect, loss of balance and muscle coordination,

cogwheel movement of the legs, slurred speech, difficulty speaking (scanning speech and slow hesitant speech), difficulty writing, difficulty swallowing, proprioceptive dysfunction, abnormal sensations (numbness, pins and needles), shortness of breath, headaches, itching, rashes, hair loss, seizures, tremors, muscular twitching or fasciculation, abnormal gait, stiffness, subnormal temperature, sensitivities to common chemicals, sleeping disorders, facial pallor, bladder and bowel problems, difficulty walking, pain, tachycardia, stroke-like episodes, food intolerances and alcohol intolerance, and partial or complete paralysis.	cogwheel movement of the legs, slurred speech, difficulty speaking (scanning speech and slow hesitant speech), difficulty writing, difficulty swallowing, proprioceptive dysfunction, abnormal sensations (numbness, pins and needles), shortness of breath, headaches, itching, rashes, hair loss, seizures, tremors, muscular twitching or fasciculation, abnormal gait, stiffness, subnormal temperature, sensitivities to common chemicals, sleeping disorders, facial pallor, bladder and bowel problems, difficulty walking, pain, tachycardia, stroke-like episodes, food intolerances and alcohol intolerance, and partial or complete paralysis.
M.S. can cause orthostatic intolerance (dizziness or faintness on standing).	M.E. commonly causes severe orthostatic intolerance (which often worsens to become severe Postural Orthostatic Tachycardia Syndrome and/or Neurally Mediated Hypotension).
Short-term memory loss and other forms of cognitive impairment occur in 50% of M.S. patients. 10% of M.S. patients have cognitive impairments severe enough to significantly affect daily life.	Short-term memory loss and other forms of cognitive impairment occur in 100% of M.E. patients. Almost all M.E. patients have cognitive impairments that significantly affect daily life.
M.S. patients often become much more ill in even mildly warm weather. Cold weather can also cause significant problems.	M.E. patients often become much more ill in even mildly warm weather. Cold weather can also cause significant problems.
M.S. is thought to cause a breakdown of the blood brain barrier.	M.E. is thought to cause a breakdown of the blood brain barrier.
M.S. can affect autonomic nervous system function (including involuntary functions such as digestion and heart rhythms).	M.E. can affect autonomic nervous system function (including involuntary functions such as digestion and heart rhythms).
A positive Babinski's reflex is consistent with several neurological conditions, including M.S.	A positive Babinski's reflex (or extensor plantar reflex) is consistent with M.E.
The Romberg test will often be abnormal in M.S. (This test measures neurological dysfunction.)	The Romberg test will be abnormal in 95% or more of M.E. patients.
An abnormal neurological exam is usual in M.S. Abnormalities are also commonly seen in neuropsychological testing in M.S.	An abnormal neurological exam is usual in M.E. Abnormalities are also commonly seen in neuropsychological testing in M.E.
M.S. causes a certain type of brain lesion detectable in MRI brain scans. Abnormalities are also seen in EEG and QEEG brain maps and SPECT brain scans in M.S.	M.E. causes a certain type of brain lesion detectable in MRI brain scans. Abnormalities are also seen in EEG and QEEG brain maps and SPECT brain scans in M.E.
Hypothyroidism is found in many M.S. patients.	Hypothyroidism is found in almost all M.E. patients.
The glucose tolerance test is often abnormal in M.S.	The glucose tolerance test is often abnormal in M.E.

Low blood pressure readings (usually low-normal) are common in M.S.	Low blood pressure readings are extremely common in M.E. Severely low blood pressure readings as low as, or lower than, 84/48 are common in severe M.E. or those having severe relapses. This can occur at rest or as a result of orthostatic or physical overexertion. Circulating blood volume measurements of only 50% to 75% of expected are also commonly seen in M.E.
Patients with M.S. have an increased risk of dying from heart disease or vascular diseases.	Deaths from cardiac problems are one of the most common causes of death in M.E.
Although M.S. is primarily neurological, it also has aspects of autoimmune disease.	Although M.E. is primarily neurological, it also has aspects of autoimmune disease.
M.S. usually affects people between the ages of 20 and 40 years, and the average age of onset is approximately 34 years. Onset occurs between the ages of 20 to 40 years in 70% of patients.	The average ages affected by M.E. are similar to those seen in M.S. However, the average age of *onset* may be significantly younger in M.E.
M.S. was once thought to be rare in children, but we now know that around 5% of M.S. sufferers are under 18.	Around 10% of M.E. sufferers are under 18.
M.S. affects more than a million adults and children worldwide.	M.E. affects more than a million adults and children worldwide.

As well as there being many similarities in symptoms, the brain scans from M.E. and M.S. patients are often very similar. M.S. and M.E. both cause a certain type of brain lesion detectable in brain scans. Those with M.S. tend to have fewer brain lesions of a larger size, while M.E. is associated with a greater number of these lesions of a somewhat smaller size.

M.E. and M.S. are so similar medically that they are sometimes misdiagnosed as one another.

The names used for M.E. and M.S. also indicate the similarities between the two diseases. M.S. was first described in 1868, and M.S. has also been known as 'disseminated sclerosis' or 'encephalomyelitis disseminate.' Myalgic Encephalomyelitis has existed for centuries but was first comprehensively scientifically documented in 1934, when an outbreak of what at first seemed to be Poliomyelitis (Polio) occurred in Los Angeles (M.E. occurs in outbreaks as well as sporadically).

The term Myalgic Encephalomyelitis was coined in 1956. Earlier names for M.E. include 'atypical Polio' and atypical Multiple Sclerosis.'

Both M.S. and M.E. have been correctly classified as organic diseases of the central nervous system in the World Health Organization's International Classification of Diseases for many decades. M.S. is classified at G 35 and M.E. at G 93.3.

Why are people with M.E. often envious of people with M.S.?

M.S. and M.E. are distinct diseases, but they are in many ways very similar medically.

However, despite the medical similarities, the two diseases are treated very differently politically and socially.

The differences between the political and social treatment of M.S. and M.E. are the reason for M.E. patients' envy of M.S. sufferers.

Multiple Sclerosis	Myalgic Encephalomyelitis
M.S. is a neurological disease, so M.S. patients are treated primarily by neurologists. In countries such as Australia, Canada, New Zealand, the USA and the UK, the majority of M.S. patients have access to a neurologist who is knowledgeable about M.S.	M.E. is also a neurological disease that is appropriately treated by a neurologist, yet very few M.E. patients have access to a doctor who knows even the most basic facts of M.E., let alone access to a neurologist who has experience and knowledge of M.E. The vast majority of M.E. patients have no access to appropriate medical care at all.
Media reports on M.S. are of a high standard. If reporters put out stories about M.S. that were not factual, there would be a public outcry and then an apology made.	Media reports on M.E. are of a very low standard. It is extremely common to read articles claiming to be about M.E. but which do not contain even one accurate fact about the disease. Complaints made by M.E. patients and experts are ignored.
Media reports on those who have experienced some recovery from M.S. involve genuine M.S. patients.	Media reports of 'miracle recoveries' from M.E. touting one pseudo-treatment or another are very common. However, the patients described did not have M.E. (or any other serious neurological disease).
M.S. advocacy groups do good work for M.S. patients and help raise awareness and funds for research. M.S. groups are run for and by M.S. patients.	The vast majority of M.E. advocacy groups do not advocate on behalf of M.E. patients but instead work directly against the best interests of M.E. patients. The vast majority of these groups are **not** run for or by M.E. patients and their agendas are **not** helping M.E. patients. These groups often distribute information on M.E. which is completely inaccurate and which also belittles and misrepresents M.E. patients.
M.S. charities would never support treatments for M.S. which had zero chance of success, and which very often caused a severe and prolonged deterioration of the disease, or even death.	So-called M.E. charities very often fully support and push 'treatments' for M.E. which have zero chance of success, and which very often cause a severe and prolonged deterioration of the disease, or even death.
M.S. is a well-known illness, and patients are generally treated appropriately by doctors and other medical staff.	M.E. is an illness that most medical staff are not well educated about. M.E. patients are often treated inappropriately by doctors and other medical staff.
People with M.S. will generally qualify for the welfare and medical insurance payouts they are entitled to.	People with M.E. are often denied the appropriate welfare and medical insurance payouts they are entitled to.
M.S. receives many millions of dollars in government funding for research, and millions more are raised each year by M.S. charities around the world.	M.E. receives no government funding worldwide and very little is raised by charities. What little is raised by these groups is virtually always spent researching non-M.E. patient groups or mixed patient-groups, but even those studies which do include a small proportion of M.E. patients are useless, because mixed patient groups make any results meaningless.

When research says it involves M.S. patients one can have a high degree of confidence that this is indeed the case.	When research says it involves M.E. patients one can only have a low degree of confidence that this is indeed the case.
When a patient with M.S. chooses euthanasia, public sympathy is expressed for the degree of pain and suffering that must have led to such a choice.	When a patient with M.E. chooses euthanasia, public derision is often expressed. Even when, as is almost always the case with euthanasia, the person with M.E. was severely affected and bedbound, it is very often blithely claimed by the media either that the patient was not ill at all, or had a mild disease that could be easily cured within weeks *if the patient only truly wanted to get better.*

Although M.S. and M.E. are very similar medically, they are worlds apart politically and socially.

Why is the public perception of M.E. so different to that of M.S.?

The public perception of M.S. reflects the reality. Most members of the public are aware of the basic facts: that M.S. is a neurological disease affecting the muscles, and that it can be very disabling or fatal. Understanding of these facts is also reflected in the way the media handles M.S. and government policy on M.S.

The public perception of M.E. could not be further removed from the medical reality of the disease.

Most members of the public, if they have heard of M.E., have heard an entirely inaccurate account of the disease which they mistakenly believe to be based on science. Despite the fact that M.E. is a serious neurological disease comparable to M.S., Lupus and Polio, M.E. is seen by most of the public and even by most of the medical profession as 'trivial.' M.E. is perceived and presented similarly by most of the media and by government. M.E. patients are treated differently to those with comparable diseases such as M.S. The contrast is stark.

There is an abundance of evidence showing that M.S. is an organic neurological disease that can be severely disabling or fatal. The same is true of M.E. The evidence supporting M.E. is no less compelling, although you would not know this from the way M.E. is dealt with. If anything, M.E. has more scientific credibility; it is far easier to diagnose due to its acute onset and more obvious, systemic and unique pathology; the cause is far more certain in M.E.

In short, the reason M.E. is treated so differently to M.S., despite their being comparable diseases, has nothing to do with science or evidence, and everything to do with MONEY.

M.E. patients are being (mis)treated based purely on financial considerations. Financial vested interest groups have subverted and obscured the reality of M.E. for their own benefit. Many millions of dollars are being made (or saved) by powerful medical insurance companies, and others, by this scam. (This is explained in detail in What is M.E.?)

This lucrative fiction about M.E. is widely accepted in the community, but it has about as much to do with science as astrology has with astronomy. It has been scientifically disproven hundreds of times over, even *before* the large scale cover-up/scam was created.

Repeating a lie over and over again will never make it true, but it seems it often will make lots of people believe it to be true, especially if the sources are seen as 'authorities.' **That's why this abusive and unscientific money-making fiction about M.E. has continued for 20 years and has only become more extreme and entrenched over time.**

M.S. is not being targeted in the same way as M.E. by insurance companies etc. This is a matter of timing: M.S. emerged earlier, received more medical attention and has been longer established within mainstream medicine than M.E. Because of this fortunate timing difference, M.S. has escaped the modern manipulation for profit which has plagued M.E. for the last 20 years.

For example, doctors working for medical insurance companies are able to get influential government advisory positions in the field of health which play a large role in determining how diseases are treated, categorized and defined. Giving corporations with vested interests the power to unscientifically re-define and/or re-classify (i.e. wrongly re-classify) a disease to suit their own interests can be *immensely* lucrative for them. Political interests have determined how M.E. is dealt with and how it is perceived, which is not true of M.S.

The reason M.E. patients are so poorly and inappropriately treated is clear. How to stop this abuse, when governments, so-called M.E. charities and the media are colluding in a cover-up for their own benefit, is far less clear.

How patients with M.E. can work to change the situation when they are so ill and disabled, and when so many are too ill to even be able to read the facts about what is happening, and when they have so little other support, is not clear. How can patients with M.E. get through to the vast majority of the public who refuses to believe government and industry could be so immoral (despite ample examples of past transgressions)? How can patients with M.E. convince others of the truth when so many seemingly benign companies, government departments, journalists or supposedly patient-based organisations are producing so much mutually supportive and superficially convincing propaganda?

These are hard questions and simply enormous problems which M.E. patients are forced to deal with, and which M.S. patients need not ever consider.

M.E. patients sometimes wonder how their lives would be different if they instead had M.S....

People with M.E. wonder if those with M.S. know how lucky they are to be able to go to the emergency room when they are very unwell, and in fear of dying, and know that they will be treated with respect and given the appropriate care; rather than laughed at, mocked in front of the other patients, refused tests or treatment and just sent home. M.E. sufferers wonder if M.S. patients know how lucky they are that millions of dollars are being spent trying to cure their disease. That knowledge must be so comforting.

People with M.E. wonder if those with M.S. know how lucky they are to have access to a doctor who knows at least the basic facts of their disease. Very few patients with M.E. have such a 'luxury.' Some M.E. patients wonder if M.S. patients know how lucky they are to not have to worry that the latest unscientific study or article that claims to be about their disease will cause those around them to mistreat them due to the study involving an unrelated patient group, which suits the authors own vested interests.

M.S. can be very severe. So can M.E. Being severely ill is hard, but being severely ill through mistreatment, apathy and neglect – as is the case with many severe M.E. patients – is even harder to deal with.

Severe injury (or death) is inflicted on thousands of people with M.E. every year because of inappropriate medical advice, but this does not seem to cause any public concern. There would be outrage if even a tiny fraction of the harm done to M.E. patients was done to people with other diseases, but the outrage is just not there for us. For M.E. patients *that* is very, very hard to live with.

It is also hard for M.E. patients to live with the fact that some people with M.E. are reduced to poverty by refusal of welfare or insurance payments, which they would have received if they had M.S. That some people with M.E. have died from inappropriate and cruel medical mistreatment, and their abusers will never be brought to justice. That some M.E. patients have had their family and friends disown them due to misconceptions about M.E. That some M.E. patients have their rights taken away, and subjected to treatments that cannot improve their condition but which carry an enormous risk of worsening the disease seriously, or causing death. That some parents of children with M.E. have been charged with causing their child's illness (falsely accused of Munchausen by Proxy) and had their children removed from their care and then seriously medically abused. That small children very ill with M.E. have been thrown into swimming pools (and very nearly drowned), or denied food or contact with family, in an attempt to force them to do things they are too ill to do because of their disease.

M.E. is one of the most severely disabling and devastating diseases there is. Yet despite all the medical advances in today's high-tech world, it is as though M.E. patients live in another era and receive only the most primitive and rudimentary care – if indeed they receive any care at all.

Conclusion

People will often say to M.E. patients 'at least you don't have M.S. It could be worse, you should be grateful.' But if anything the opposite is true. Taking everything into account, the physical reality of each disease plus the misconceptions and mistreatment associated with M.E., it is very hard to see how *anyone* would ever choose M.E. over M.S., if such a choice were possible.

That isn't to say that M.S. isn't a terrible disease, or that those with M.S. have any more resources or funding than they rightly deserve and need, or that everyone with M.S. always gets every service they need easily and will always have a very supportive family. *Of course not.* The point is that it makes no sense that patients with these two very similar diseases are treated so differently just because of political manipulation for profit; that scientific reality, ethics and logic count for so little.

It is as bizarre and unfair as if those with broken arms were given x-rays and had the broken bone set and put in a cast until it had healed, while those with broken legs were told to go home and stop wasting the doctor's time, or that perhaps taking up jogging would make them feel better.

M.S. is a disabling neurological disease that causes a high degree of suffering. So is M.E. However, there is a whole other world of suffering experienced by M.E. patients which is unknown to M.S. patients and others with diseases where the public perception and political treatment of the disease is closely aligned with the medical reality. It is an additional type of suffering which can be as much a burden as the disease itself. When you combine these political problems with a disease as serious as M.E., it makes M.E. hell on earth.

The treatment of people with M.E. must be based on science at last, as is the treatment of M.S. patients. All M.E. patients want is to be treated the same way as those with M.S. and other comparable illnesses. All M.E. patients want is for studies on M.E. to actually involve M.E. patients, for the term M.E. to only be used to describe actual M.E. patients, for the facts about M.E. to be taught at medical schools in the same way M.S. facts are, for appropriate money to be made available for M.E. research, for government policy on M.E. to reflect the reality of M.E., and for the media (including medical journals) to write articles about M.E. with the same standard of factual accuracy as articles on M.S., and other diseases.

These things don't seem much to ask for in this day and age, but right now, they out of reach for M.E. patients and if anything they get further and further away as each year passes.

More information and additional notes on this text

- For more information on this topic, including a table describing the medical *differences* between M.S. and M.E., please see the full-length version of this text on the HFME website. Please also note that none of these charts is designed to be comprehensive or detailed enough to be used to differentiate between an M.S. or M.E. diagnosis. Additional resources on Multiple Sclerosis used in creating this paper are listed in the full-length version of this text on the website.

- *An important note:* To be clear, while M.E. can be far more disabling than M.S. and some medical aspects of it are worse than is seen in M.S., there is no doubt that *some aspects* of M.S. are *undoubtedly worse* than *some aspects* of M.E. The level of suffering can be very high in both diseases, and nobody with either disease has it easy. Those with the worst deal illness-wise are those with the most severe forms of either M.E. or M.S.

  To read this paper and feel that the statement being made is that M.E. is always worse than M.S., and that M.S. does not cause immense suffering or doesn't deserve more funding for support, is to miss the point of this paper entirely.

  People with M.E. and M.S. have been dealt a very cruel blow and need and deserve all the support and kindness they can get. Both are absolutely devastating diseases.

To read a fully-referenced version of the medical information in this text compiled using information from the world's leading M.E. experts, please see the 'What is M.E.?' paper on page 113 of this book or on the HFME website.

Acknowledgments
Thanks to Lesley Ben for editing this paper.

Relevant quotes

'I have a friend who has M.S. and she is really very independent and able to get about (unlike me). My Dr isn't very helpful but at least she is kind. My friend with M.S. has said how similar our illnesses are, yet she has a special M.S. nurse, a supportive Dr and the knowledge that when she tells someone she has M.S. they will be understanding and non-judgemental.

If only we could have half of that I would be happy. I had to use a wheelchair for 2 years and I had people saying I was lazy and why was I 'carrying on' like that! Can you believe it?! When I was so severely ill and couldn't get out of bed for months on end people told my husband I needed 'motivating' and that I was probably having a nervous breakdown and was depressed... If only I had M.S.!!!

I really do feel that we are left to cope with such a debiltating illness alone almost. My husband has nearly lost his job because of time off he has had to take to care for me when I'm at my worst. We struggle financially as we only have one wage coming in and yet we get no help or support from anywhere, and yet if I was suffering from M.S. people would be appalled at my situation. As it is, most people don't even think I'm ill and that I should just 'pull myself together' - if only I could! And this is the joke, we are too _____ ill to stand up for our rights and make a change!'
LENA, M.E. PATIENT

'i have to admit i get very resentful when i see the adverts on television for everything except M.E. (not to deny serious diseases their rightful place in public awareness, but it's hard to be left out in the cold year after year, decade after decade). on a bad day i probably feel as wretched as people with severe M.S.; on a "good" day i probably feel about as well as people with a mild case of M.S. but then it's probably the same for them ... degrees of severity.

having struggled with M.E. in the face of doubt, invalidation, "pep talks", psychologizing, and so on since i was 28 ... i'll be turning 55 next month ... i think that had someone come to me and given me the choice, i'd have chosen M.S. for exactly the reasons put forth by so many others. that M.E. isn't recognised as "real" (despite the mountains of evidence to the contrary) often pushes me to the brink of absolute despair, and seeing so many other people receiving the kind of care i need just adds insult to injury.

i have long since lost count of the times i've been scolded and sent home untreated, only to return a few days later with symptoms no one can deny or ignore. I fully expect my cause of death will be medical neglect.'
NAMASTE, SEVERE M.E. PATIENT

'My sister who is 34 has known since August 2008 that she has M.S. She was paralyzed on 1 side of her body as she went to the hospital. There she got the diagnosis M.S. She got antibiotics by a drip and got well again. Now she has no complaints except for some tiredness. She now works 4 days a week instead of 5. As her neurologist heard from her about me, he wanted to see me to do some tests because of my neurological problems. I got an MRI, but it did not indicate M.S. Later on he told my sister that she had to do less and I had to do more!!!!! He had seen me twice and didn't know much about me or about M.E. My sister gets all the attention and understanding. From family, from friends, from work, from doctors and she probably won't have any problem getting care and a wheelchair and stuff when she may need it later on. I had to buy my wheelchair by myself as I improved somewhat to be able to use it finally. There are wheelchairs in which you can lay down. I've asked for one, but I didn't get one. As I was a bit better and could walk a few meters again I had to give back my special electric wheelchair in which you can lay down. I also get only 1 hour of ADL-care a day while I'm bedridden for 10 years now. I also haven't seen my family for 10 years, because they think I'm lazy and don't want to work.

I've lost all my teeth at age 36, because my gp didn't believe I had severe burning acid.

There are months I can't swallow, but I don't get tube-feeding. Not even fluid by IV. I wish I had M.S. As I heard of my sister having M.S. I was jealous and I still am. I so hoped I had M.S. as I went back to the neurologist for the result of the MRI. When I got home I cried in frustration and disappointment. I still feel very sad and disappointed about it weeks later.'
INGEBORG GEUIJEN, M.E. PATIENT (NETHERLANDS), AUTHOR OF WWW.BORGOFSPACE.COM

'I do think that the simultaneous rise of AIDS at the same time as the 80's US M.E. epidemics sucked the life out of any possible attention by public health towards M.E. It is so hard to talk about because AIDS is a TERRIBLE disease but they got the funding and the recognition and research because it was a very powerful, large, population group here in NA, well used to activism, and by the way, men, at least in the early days. And just look what has happened for PWA's, all the research and progress, all the social support and awareness. There's quite a strong AIDS group here. I found out that a woman I know with it, who has a husband and kids, who functions as healthy running around all over the place, and who gets substantial caregiver hours. And here I am, unable to do my own shopping or housework and can't get one lousy hour. And one can never talk about it, it's not PC <sigh>. I do not begrudge them their care etc. but only wish for 1/10th of the help that they get (or that those with M.S. get).'
AYLWIN CATCHPOLE, M.E. PATIENT 20+ YEARS (CANADA)

'I've just read your M.E. vs M.S. page. 20 yrs ago I was treated and tested by a neurologist who thought I had M.S. When I was diagnosed with M.E. instead of M.S., I was thought I was fortunate, but often since then I've wished the M.S. diagnosis was the correct one instead of M.E. Your article expresses my feelings completely. Thankyou.'
WENDY, M.E. PATIENT

'I actually walked away from mainstream medical care about 15 yrs ago out of utter frustration. Now I am having another "go" just to try and obtain any help that might be had. And my perception is that it's worse now. And my patience is running out. again. I almost wish I could be MISdiagnosed with M.S. or other, more respected disease, just to get better treatment.'
AYLWIN CATCHPOLE, M.E. PATIENT 20+ YEARS (CANADA)

'Hate to say it, but the only folks I know who have M.S. worse than we have M.E. are two women with the galloping progressive rapid kind, who are in wheelchairs & care, but they won't live much longer. But everyone else I've ever known with M.S. can run circles around me!! No offense to anybody, I have known people with M.S., Lupus, Lyme, AIDS (not just asymptomatic HIV) and all kinds of other similar conditions, and to a one they can all run circles around me...until it's time to die that is. (Then they are more disabled than us, but only then). <sigh>'
AYLWIN CATCHPOLE, M.E. PATIENT 20+ YEARS (CANADA)

'I have had M.S. for 10 years and I have been amazed by the similarities to people I know with M.E. and I fully sympathise with the issues regarding the different attitude of the public and professionals to the two illnesses and about the different levels of support that are available. In a way I think it is easier that people with M.S. are not expected to get better. Over time I have come to terms with the gradual deterioration and I don't look for "cures", etc. and although it can be annoying when people constantly TELL me that I look well, and therefore assume I have no symptoms at that time, I think it must be much worse for M.E. sufferers to have so many people expect them to get better and to keep asking them about their progress. Thank you for the article and I wish everyone with M.S. or M.E. the courage to deal with each day.'
PAM, M.S. PATIENT

'I have been, and continue to be, bullied by ignorant doctors. My husband didn't believe how bad it was until he started to come with me. Now he says "I can't believe how they treat you". It feels good to know it's not just me now. Even my ex thought I was just overreacting when I would tell him about how doctors treated me. What really upsets me is I have a friend who has M.S. yet can still work etc she gets immunoglobulin injections and the doc treats her amazingly. It makes me so hurt and angry that because I have M.E. I am treated with such disrespect and contempt. Sorry, I havr had that bottlerd up for a long time and until now., had nowhere to say it where it would be understood.'
CAROLINE, SEVERE M.E. PATIENT

'Just found this site today. I have M.E. and my sister has M.S. For a long time I have been amazed at how similar our symptoms are but I am amazed to see it for fact. I have had very little medical support (I've been diagnosed for 19 yrs). She has been diagnosed for 5 yrs and has a great source of medical support. It really should be more equal - well hopefully the future will be a more supportive one for M.E. The truth still stands that we would both rather not need support but have good health. To everyone with either illness, support each other and laugh when you can.'
M.E. PATIENT

'I don't think there actually are any other disease-sufferers getting as little help as we do. And yet there are

very few diseases (if any), which impact, destroy and restrict your life and ability to function as much as M.E. does. I would not get any caregiver hours or assistive devices either, if it was not for the fact that I also have a connective tissue disorder called the Ehlers-Danlos syndrome (EDS). EDS is not the actual reason for my disability and need of help. The reason for me needing help and being so sick and disable is M.E. But with "only" M.E. I would not get ANY kind of help. Regardless of the fact that I would not stay alive without any caregiver hours.'
M, SEVERE M.E. PATIENT

'It has become obvious to me that we are dealing with both a vasculitis and a change in vascular physiology. Numerous other physicians have supported this finding. The recent interpretation of the cause of Multiple Sclerosis (M.S.) as an injury of the microvasculization causing the injury of the schwann cells that in turn causes the demyelination injuries of M.S. has been added to that of paralytic poliomyelitis as an essential vascular injury. Paralytic poliomyelitis was thought to be a primary injury to the anterior horn cells of the spinal cord but is now recognized as a vasculitis injuring the circulation to the anterior horn cells. Poliomyelitis is generally a non-progressive, specific site injury, although post-polio syndrome has challenged that belief. M.S. is a recurrent more fulminant physiological vascular injury.

M.E. appears to be in this same family of diseases as paralytic polio and M.S. M.E. is less fulminant than M.S. but more generalized. M.E. is less fulminant but more generalized than poliomyelitis. This relationship of M.E.-like illness to poliomyelitis is not new and is of course the reason that Alexander Gilliam, in his analysis of the Los Angeles County General Hospital M.E. epidemic in 1934, called M.E. atypical poliomyelitis.'
DR BYRON HYDE 2006

'The term myalgic encephalomyelitis (means muscle pain, my-algic, with inflammation of the brain and spinal cord, encephalo-myel-itis, brain spinal cord inflammation) was first coined by Ramsay and Richardson and has been included by the World Health Organisation (WHO) in their International Classification of Diseases (ICD), since 1969. It cannot be emphasised too strongly that this recognition emerged from meticulous clinical observation and examination.'
PROFESSOR MALCOLM HOOPER 2006

'There is ample evidence that M.E. is primarily a neurological illness. It is classified as such under the WHO international classification of diseases (ICD 10, 1992) although non neurological complications affecting the liver, cardiac and skeletal muscle, endocrine and lymphoid tissues are also recognised. Apart from secondary infection, the commonest causes of relapse in this illness are physical or mental over exertion.'
DR ELIZABETH DOWSETT

'Possible costing for M.E. support has been based on 3 times the cost of maintenance for Multiple Sclerosis on the supposition that M.E. is [up to] 3 times as common. The only costs that we can be sure of are those derived from the failure of appropriate management, and of inappropriate assessments which waste vast sums of money and medical time while allowing patients to deteriorate unnecessarily. Research workers must be encouraged and appropriately funded to work in this field. However they should first be directed to papers published before 1988, the time at which all specialised experience about poliomyelitis and associated infections seem to have vanished mysteriously!'
DR ELIZABETH DOWSETT

'People in positions of power are misusing that power against sick people and are using it to further their own vested interests. No-one in authority is listening, at least not until they themselves or their own family join the ranks of the persecuted, when they too come up against a wall of utter indifference.'
PROFESSOR M. HOOPER 2003

There are many clinical and laboratory similarities in M.E. and M.S., but what separates them is: the plethora of systemic manifestations in M.E., the orthostatic tachycardia seen in M.E., the outbreaks of M.E., the striking involvement of muscle in M.E. and the muscle pathology seen in M.E., the characteristic myalgias and arthralgias in M.E., and the symptoms such as cold extremities and flu-like symptoms etc. seen in M.E. These features are not seen in M.S. and their presence may even preclude a M.S. diagnosis.
FROM CHARLES M POSER MD IN THE BOOK THE CLINICAL AND SCIENTIFIC BASIS OF MYALGIC ENCEPHALOMYELITIS (PARAPHRASED BY THE AUTHOR)

The HUMMINGBIRDS' FOUNDATION for M.E. (HFME)
Fighting for the recognition of Myalgic Encephalomyelitis based on the available scientific evidence, and for
patients worldwide to be treated appropriately and accorded the same basic human rights as those with similar
disabling and potentially fatal neurological diseases such as Multiple Sclerosis.

What is M.E.? A historical, medical and political overview

 Myalgic Encephalomyelitis (M.E.) is a debilitating acquired neurological disease which has been recognised by the World Health Organisation (WHO) since 1969 as a distinct organic neurological disorder.

M.E. can occur in both epidemic and sporadic forms, and over 60 outbreaks of M.E. have been recorded worldwide since 1934. M.E. is similar in a number of significant ways to Multiple Sclerosis, Lupus and Poliomyelitis (Polio).

M.E. can be extremely severe and disabling and in some cases the disease is fatal.

Is M.E. a new illness? What does the name Myalgic Encephalomyelitis mean?

The disease we now know as Myalgic Encephalomyelitis is not a new disease. M.E. is thought to have existed for centuries (Hyde 1998, [Online]) (Dowsett 1999a, [Online]).

In 1956 the name Myalgic Encephalomyelitis was created. The term was invented jointly by Dr A Melvin Ramsay, who coined this name in relation to the Royal Free Hospital epidemics that occurred in London in 1955 – 1957, and by Dr John Richardson, who observed the same type of illness in his rural practice in Newcastle-upon-Tyne during the same period. It was obvious to these physicians that they were dealing with the consequences of an epidemic and endemic infectious neurological disease (Hyde 1998, [Online]) (Hyde 2006, [Online]).

The term Myalgic Encephalomyelitis means: My = muscle, algic = pain, encephalo = brain, mye = spinal cord, itis = inflammation (Hyde 2006, [Online]).

As M.E. expert Dr Byron Hyde writes:

> The reason why these physicians were so sure that they were dealing with an inflammatory illness of the brain is that they examined patients in both epidemic and endemic situations with this curious diffuse brain injury. In the epidemic situation with patients falling acutely ill and in some cases dying, autopsies were performed and the diffuse inflammatory brain changes are on record (2006, [Online]).

The Wallis description of M.E. was created in 1957, and in 1959 Sir Donald Acheson (a former UK Chief Medical Officer) conducted a major review of M.E.

In 1962 the distinguished neurologist Lord Brain included M.E. in the standard textbook of neurology. In recognition of the large body of compelling research that was available, M.E. was formally classified as an organic disease of the central nervous system in the World Health Organisation's International Classification of Diseases in 1969.

In 1978 the Royal Society of Medicine held a symposium on Myalgic Encephalomyelitis at which M.E. was accepted as a distinct entity. The symposium proceedings were published in The Postgraduate Medical Journal later that same year. The Ramsay case description of M.E. was published in 1981 (Hooper et al. 2001, [Online]).

Since 1956 the term Myalgic Encephalomyelitis has been used to describe the illness in the UK, Europe Canada and Australasia. This term has stood the test of time for more than 50 years. The recorded medical history of M.E. as a debilitating organic neurological illness affecting children and adults is substantial; it spans over 80 years and has been published in prestigious peer-reviewed journals all over the world (Hyde 1998, [Online]) (Hooper 2003a, [Online]) (Dowsett 2001b, [Online]).

As award winning microbiologist and M.E. expert Dr Elizabeth Dowsett explains: 'There is ample evidence that M.E. is primarily a neurological illness, although non-neurological complications affecting the liver, cardiac and skeletal muscle, endocrine and lymphoid tissues are also recognised' (n.d.b, [Online]).

M.E. is not defined by mere 'fatigue'

M.E. is not synonymous with being tired all the time. If a person is very fatigued for an extended period of time this does not mean they are having a 'bout' of M.E. Such a suggestion is no less absurd than to say that prolonged fatigue means a person is having a 'bout' of Multiple Sclerosis, Parkinson's disease or Lupus. If a person is constantly fatigued this should not be taken to mean that they have M.E., no matter how severe or prolonged their fatigue is.

Fatigue is a symptom of many different illnesses as well as a feature of normal everyday life – but it is not a defining symptom of M.E., or even an essential symptom of M.E. The terms 'fatigue' and 'chronic fatigue' were not associated with defining this illness until the entity of 'Chronic Fatigue Syndrome' was created in 1988 in the USA (Hyde 2006, [online]). But M.E. and 'CFS' are *not* synonymous terms.

'Fatigue' and 'feeling tired all the time' are not at all the same thing as the very specific type of *paralytic muscle weakness* or *muscle fatigue* which *is* characteristic of M.E. (caused by mitochondrial dysfunction) and which affects every organ and cell in the body, including the brain and the heart. This causes – or significantly contributes to – such problems in M.E. as cardiac insufficiency (a type of heart failure), orthostatic intolerance or POTS (inability to maintain an upright posture), blackouts, reduced circulating blood volume (and pooling of the blood in the extremities), seizures (and other neurological phenomena), memory loss, problems chewing/swallowing, episodes of partial or total paralysis, muscle spasms/twitching, extreme pain, problems with digestion, vision disturbances, and breathing difficulties.

These problems are exacerbated by even trivial levels of physical and cognitive activity, sensory input and orthostatic stress beyond a patient's individual limits. People with M.E. are made very ill and disabled by this problem with their cells; it affects virtually every bodily system and has also lead to death in some cases. Many patients are housebound and bedbound and are often so ill that they feel they are about to die. People with M.E. would give *anything* to only be severely 'fatigued' or 'tired all the time' (Bassett 2010, [Online]).

Fatigue, post-exertional fatigue or malaise may occur in many different illnesses such as various post-viral fatigue states or syndromes, Fibromyalgia, Lyme disease, and many others, but what is happening with M.E. patients is an entirely different and unique problem of a much greater magnitude. These terms are not accurate or specific enough to describe what is happening in M.E.

M.E. is a neurological illness of extraordinarily incapacitating dimensions that affects virtually every bodily system – not a problem of 'chronic fatigue' (Hyde 2006, [Online]) (Hooper 2006, [Online]) (Hooper & Marshall 2005a, [Online]) (Hyde 2003, [Online]) (Dowsett 2001, [Online]) (Hooper et al. 2001, [Online]) (Dowsett 2000, [Online]) (Dowsett 1999a, 1999b, [Online]) (Dowsett 1996, p. 167) (Dowsett et al. 1990, pp. 285-291) (Dowsett n.d., [Online]).

- For more information see M.E. is not fatigue, or 'CFS'. Many of the world's leading M.E. experts have spoken out strongly against claims that 'fatigue' is the defining/essential symptom of M.E. See M.E. is not defined by 'fatigue' to read some of their comments. For more information on the symptoms of M.E., including the unique reaction people with M.E. have to activity, see: The ultra-comprehensive M.E. symptom list.

If M.E. and 'CFS' are not synonymous terms, why do some groups claim that they are? What is 'CFS'?

The disease category of 'CFS' was created in a response to an outbreak of what was unmistakably M.E., but this new name and definition did not describe the known signs, symptoms, history and pathology of M.E. It described a disease process that did not, and could not exist.

Why was the renaming and redefining of the distinct neurological disease M.E. allowed to become so muddied? Indeed, why did Myalgic Encephalomyelitis suddenly need to be renamed or redefined at all?

The answer is money. There was an enormous rise in the reported incidence of M.E. in the late 1970s and 1980s, alarming medical insurance companies in the US. So it was at this time that certain psychiatrists and others involved in the medical insurance industry (on both sides of the Atlantic) began their campaign to reclassify M.E. as a psychological or 'personality' disorder, in order to side-step the financial responsibility of so many new claims (Marshall & Williams 2005a, [Online]).

As Professor Malcolm Hooper explains:

> In the 1980s in the US (where there is no NHS and most of the costs of health care are borne by insurance companies), the incidence of M.E. escalated rapidly, so a political decision was taken to rename M.E. as "chronic fatigue syndrome", the cardinal feature of which was to be chronic or ongoing "fatigue", a symptom so universal that any insurance claim based on "tiredness" could be expediently denied. The new case definition bore little relation to M.E.: objections were raised by experienced international clinicians and medical scientists, but all objections were ignored… To the serious disadvantage of patients, these psychiatrists have propagated untruths and falsehoods about the disorder to the medical, legal, insurance and media communities, as well as to government Ministers and to Members of Parliament, resulting in the withdrawal and erosion of both social and financial support [for M.E. patients]. Influenced by these psychiatrists, government bodies around the world have continued to propagate the same falsehoods with the result that patients are left without any hope of understanding or of health service provision or delivery. As a consequence, government funding into the biomedical aspects of the disorder is non-existent (2003a, [Online]) (2001, [Online]).

The psychiatrist Simon Wessely – arguably the most powerful and prolific author of papers which claim that M.E. is merely a psychological problem of 'fatigue' – began his rise to prominence in the UK at the same time the first CFS definition was being created in the USA (1988). Wessely, and his like-minded colleagues – a small group made up mostly but not exclusively of psychiatrists (colloquially known as the 'Wessely School') has gained dominance in the field of M.E. in the UK (and increasingly around the world) by producing vast numbers of papers which purport to be about M.E.

Wessely claims to specialise in M.E. but uses the term interchangeably with chronic fatigue, fatigue or tiredness, plus terms such as neurasthenia, CFS and 'CFS/ME' (a confusing and misleading term he created himself). He claims that psychiatric states of ongoing fatigue and the distinct neurological disorder M.E. are synonymous. Despite all the existing contradictory evidence, Wessely (and members of the Wessely School) assert that M.E. is a behavioural disorder, with no physical signs of illness or abnormalities on testing, that is perpetuated by 'aberrant illness beliefs' 'the misattribution of normal bodily sensations,' and that patients 'seek and obtain secondary gain by adopting the sick role' (Hooper & Marshall 2005a, [Online]).

The Wessely School and collaborators have assiduously attempted to obliterate recorded medical history of M.E. even though the existing evidence and studies were published in prestigious peer-reviewed journals and span over 70 years. Wessely's claims, and those of his colleagues around the world, have flooded the worldwide literature to the extent that medical journals rarely contain any factual and unbiased information about M.E. Most clinicians are effectively being deprived of the opportunity to obtain even the most basic facts about the illness.

For at least a decade, serious questions have been raised in international medical journals about possible scientific misconduct and flawed methodology in the work of Wessely and his colleagues. It is only relatively recently however that his long-term involvement as medical adviser – and board member – to a number of commercial bodies with a vested interest in how M.E. is managed have been exposed.

This is the sole reason the myth that M.E. is a psychiatric or behavioural 'fatiguing' disorder or even an 'aberrant belief system' continues: not because there is good scientific evidence for the theory, or because the evidence proving organic causes and effects is lacking, but because such a theory is so **financially and politically convenient and profitable** on such a large scale to a number of extremely powerful corporations (Hooper et al 2001, [Online]).

As Dr Elizabeth Dowsett observes, these financially motivated theories bear as much relation to legitimate science as astrology does to astronomy (1999b [Online]). Professor Malcolm Hooper goes on to explain:

> Increasingly, it is now "policy-makers" and Government advisers, not experienced clinicians, who determine how a disorder is classified and managed in the NHS: the determination of an illness classification and the provision of policy-driven "management" is a very profitable business. To the detriment of the sick, the

deciding factor governing policies on medical research and on the management and treatment of patients is increasingly determined not by medical need but by economic considerations. There is a gross mismatch between the severity and complexity of M.E. and the medical and public perception of the disorder (2003a, [Online]).

Members of the 'Wessely school' in the UK including Wessely, Sharpe, Cleare and White, their US counterparts Reeves, Straus etc of the CDC, in Australia Lloyd, Hickie etc and the clinicians of the Nijmegen group in the Netherlands each support a bogus psychiatric or behavioural paradigm of 'CFS' and recommend rehabilitation-based approaches such as cognitive behavioural therapy (CBT) and graded exercise therapy (GET) as the most useful interventions for 'CFS' patients.

It is important to be aware that none of these groups is studying patients with M.E. Each of these groups uses a definition of 'CFS,' or has created their own, which does not select those with M.E. but instead selects those with various types of psychiatric and non-psychiatric fatigue. These inappropriate interventions are at best useless and at worst extremely harmful or fatal for M.E. patients.

The creation of the bogus disease category 'CFS' has been used to impose a false psychiatric paradigm of M.E. by allying it with various unrelated psychiatric fatigue states and post-viral fatigue syndromes for the benefit of various (proven) financial and political interests. The resulting 'confusion' between the distinct neurological disease M.E. and the bogus disease category of 'CFS' has caused an overwhelming additional burden of suffering for those who suffer from M.E. and their families.

It's a huge mess, that is for certain — but it is not an *accidental* mess — that is for certain too (Hyde 2006a, [Online]) (Hooper 2006, [Online]) (Hyde 2003, [Online]) (Hooper 2003a, [Online]) (Dowsett 2001a, [Online]) (Hooper et al. 2001, [Online]) (Dowsett 2000, [Online]) (Dowsett 1999a, 1999b, [Online]).

- To read about the vast difference between M.E. and 'CFS' (and how such a small (but powerful) group of vested interest psychiatrists have come to influence the opinions of the worldwide medical community about M.E.) see: Who benefits from 'CFS' and 'ME/CFS'? and Smoke and mirrors on IIFME and also A Brief History of Myalgic Encephalomyelitis & An Irreverent History of CFS by Dr Byron Hyde
- For information on how the 'CFS' scam affects all parts of an M.E. patient's life, see M.E.: The shocking disease.

What does a diagnosis of 'CFS' actually mean?

There are now more than nine different definitions of 'CFS.' Each of these flawed 'CFS' definitions 'define' a heterogeneous (mixed) population of people with various misdiagnosed psychiatric and non-psychiatric states which have little in common but the symptom of fatigue.

The fact that a person qualifies for a diagnosis of 'CFS', based on any of the 'CFS' definitions: (a) does not mean that the patient has M.E., and (b) does not mean that the patient has any other distinct and specific illness named 'CFS.'

A diagnosis of' CFS' – based on any of the 'CFS' definitions – can only ever be a MISdiagnosis. All a diagnosis of 'CFS' actually means is that the patient has a gradual onset fatigue syndrome which is usually due to a *missed major disease*. As Dr Byron Hyde explains, the patient has:

a. Missed cardiac disease, b. Missed malignancy, c. Missed vascular disease, d. Missed brain lesion either of a vascular or space occupying lesion, e. Missed test positive rheumatologic disease, f. Missed test negative rheumatologic disease, g. Missed endocrine disease, h. Missed physiological disease, i. Missed genetic disease, j. Missed chronic infectious disease, k. Missed pharmacological or immunization induced disease, l. Missed social disease, m. Missed drug use disease or habituation, n. Missed dietary dysfunction diseases, o. Missed psychiatric disease (2006, [Online]).

Under the cover of 'CFS' certain vested interest groups have assiduously attempted to obliterate recorded medical history of M.E., even though the existing evidence has been published in prestigious peer-reviewed journals around the world and spans over 70 years. Dr Byron Hyde explains:

Do not for one minute believe that CFS is simply another name for Myalgic Encephalomyelitis. It is not. The CDC 1988 definition of CFS describes a non-existing chimera based upon inexperienced individuals who lack any historical knowledge of this disease process. The CDC definition is not a disease process. It is (a) a partial mix of infectious mononucleosis /glandular fever, (b) a mix of some of the least important aspects of M.E. and (c) what amounts to a possibly unintended psychiatric slant to an epidemic and endemic disease process of major importance. Any disease process that has major criteria, of excluding all other disease processes, is simply not a disease at all; it doesn't exist. The CFS definitions were written in such a manner that CFS becomes like a desert mirage: The closer you approach, the faster it disappears (2006, [Online]).

The only way forward for M.E. patients and all of the diverse patient groups commonly misdiagnosed with 'CFS' (both of which are denied appropriate support, diagnosis and treatment, and may also be subject to serious medical abuse) is that the bogus disease category of 'CFS' must be abandoned.

Every patient deserves the best possible opportunity for appropriate treatment for their illness and for recovery and this process must begin with a correct diagnosis if at all possible. *A correct diagnosis is half the battle won* (Hyde 2006a, 2006b, [Online]) (Hooper 2006, [Online]) (Hyde 2003, [Online]) (Hooper 2003a, [Online]) (Dowsett 2001a, [Online]) (Dowsett 2000, [Online]) (Dowsett 1999a, 1999b, [Online]) (Dowsett n.d., [Online]).

- For more information on why the bogus disease category of 'CFS' must be abandoned see: Who benefits from 'CFS' and 'ME/CFS'?, The misdiagnosis of 'CFS', Why the disease category of 'CFS' must be abandoned and Smoke and Mirrors.
- Those patients misdiagnosed with 'CFS' (and who do not have M.E.) are advised to read the following papers: The Misdiagnosis of 'CFS' and Where to after a 'CFS' (mis)diagnosis?
- *An additional note on 'fatigue':* Just as some M.E. sufferers will experience other non-essential symptoms such as vomiting or night sweats some of the time, but others will not, the same is true of fatigue. The diagnosis of M.E. is determined upon the presence of certain neurological, cognitive, cardiac, cardiovascular, immunological, endocrinological, respiratory, hormonal, muscular, gastrointestinal and other symptoms – the presence or absence of mere 'fatigue' is irrelevant.

What do the terms CFIDS, ME/CFS, CFS/ME, Myalgic Encephalopathy and ME-CFS mean?

When the terms CFS, CFIDS, ME/CFS, CFS/ME, or Myalgic Encephalopathy are used, what is being referred to may be patients with any combination of:

1. Miscellaneous psychological and non-psychological fatigue states (including somatisation disorder).

2. A self limiting post-viral fatigue state or syndrome (e.g. following glandular fever).

3. A mixed bag of unrelated, misdiagnosed illnesses (each of which features fatigue as well as a number of other common symptoms; poor sleep, headaches, muscle pain etc.) including Lyme disease, Multiple Sclerosis, Fibromyalgia, athletes over-training syndrome, depression, burnout, systemic fungal infections (Candida) and even various cancers.

4. Myalgic Encephalomyelitis patients.

The terminology is often used interchangeably, incorrectly and confusingly. However, the DEFINITIONS of M.E. and 'CFS' are very different and distinct, and it is the definitions of each of these terms which is of primary importance. *The distinction must be made between terminology and definitions.*

1. *Chronic Fatigue Syndrome* is an artificial construct created in the US in 1988 for the benefit of various political and financial vested interest groups. It is a mere diagnosis of exclusion (or wastebasket diagnosis) based on the presence of gradual or acute onset fatigue lasting at least 6 months. If tests show serious abnormalities, a person no longer qualifies for the diagnosis, as 'CFS' is 'medically unexplained.' A diagnosis of 'CFS' does not mean that a person has any distinct disease (including M.E.). The patient population diagnosed with 'CFS' is made up of people with a vast array of unrelated illnesses, or with no detectable illness. According to the latest CDC estimates, 2.54% of the population qualifies for a 'CFS' diagnosis. Every diagnosis of 'CFS' can only ever be a misdiagnosis.

2. *Myalgic Encephalomyelitis* is a systemic neurological disease initiated by a viral infection. M.E. is characterised by scientifically measurable damage to the brain, and particularly to the brain stem which results in dysfunctions and damage to almost all vital bodily systems and a loss of normal internal homeostasis.

Substantial evidence indicates that M.E. is caused by an enterovirus. The onset of M.E. is always acute and M.E. can be diagnosed within just a few weeks. M.E. is an easily recognisable distinct organic neurological disease which can be verified by objective testing. If all tests are normal, then a diagnosis of M.E. cannot be correct.

M.E. can occur in both epidemic and sporadic forms and can be extremely disabling, sometimes fatal. M.E. is a chronic/lifelong disease that has existed for centuries. It shares similarities with M.S., Lupus and Polio. There are more than 60 different neurological, cognitive, cardiac, metabolic, immunological and other M.E. symptoms. Fatigue is not a defining or even essential symptom of M.E. People with M.E. would give anything to be only 'fatigued' instead of having M.E. Far fewer than 0.5% of the population has the distinct neurological disease known since 1956 as Myalgic Encephalomyelitis.

The only thing that makes any sense is for patients with M.E. to be studied ONLY under the name Myalgic Encephalomyelitis, and for this term ONLY to be used to refer to a 100% M.E. patient group. The only correct name for this illness – M.E. as per Ramsay/Richardson/Dowsett and Hyde – is Myalgic Encephalomyelitis.

M.E. is not synonymous with 'CFS', nor is it a subgroup of 'CFS'. It is also important that the only terms which are used are those which do have an official and correct World Health Organization classification.

There is no such disease as 'CFS' – the name 'CFS' and the bogus disease category of 'CFS' must be abandoned, along with the use of other vague and misleading umbrella terms such as 'ME/CFS' 'CFS/ME' 'CFIDS, 'Myalgic Encephalopathy' and others, for the benefit of all the patient groups involved.

- For more information on why the bogus disease category of 'CFS' must be abandoned, (along with the use of other vague and misleading umbrella terms such as 'ME/CFS' 'CFS/ME' 'CFIDS' and 'Myalgic Encephalopathy' and others), see: Who benefits from 'CFS' and 'ME/CFS'?, Problems with the so-called "Fair name" campaign. Why it is in the best interests of all patient groups involved to reject and strongly oppose this misleading and counter-productive proposal to rename 'CFS' as 'ME/CFS' and Problems with the use of 'ME/CFS' by M.E. advocates, plus The misdiagnosis of CFS, Why the disease category of 'CFS' must be abandoned and Smoke and Mirrors
- *A note on the current name change proposal:* It is madness to suggest that CFS should be renamed as ME-CFS or CFS/ME or ME/CFS, as some US CFS groups are currently advocating. M.E. and CFS are not the same, only a small percentage of those (mis)diagnosed with CFS qualify for a diagnosis of authentic M.E., the vast majority do not. People with depression, Lyme disease, Candida, etc. do not need to be given an additional misdiagnosis of ME/CFS, they must instead be given a correct diagnosis finally. The fact that some of these patients, and others, may fit the Canadian criteria for 'ME/CFS' does not mean that these patients can be correctly diagnosed with M.E. – as per Ramsay/Richardson/Dowsett and Hyde – nor that these illnesses are the same or 'virtually the same' as M.E. They are not. The Canadian 'ME/CFS' Guidelines and the newer version titled the International Consensus Criteria (ICC) are not accurate M.E. definitions. They are not definitions of M.E. at all. They are both redefinitions of 'CFS' which unscientifically throw in a few facts about M.E. and by doing so unhelpfully worsen the confusion between these two very different entities. For more information see: Canadian Guidelines Review and Testing for M.E.

But isn't the name 'CFS' a big part of the problem?

The reason so many patients are ridiculed, sneered at, belittled, disbelieved, accused of exaggerating or malingering or laziness by medical staff and by friends and family members is not because of the name 'Chronic Fatigue Syndrome'! If 'CFS' had instead been given a neutral name, say 'Reeves' syndrome' or 'Holmes' syndrome,' the problems would still be exactly the same. Vested interest groups – helped in this task immeasurably by the creation of the bogus disease category of 'CFS' – would still be flooding the medical, political and media communities with lies and propaganda which could only have the end result of making patients seem utterly pathetic and undeserving of any respect or sympathy.

What else could anyone think of patients who supposedly have an illness that is mild and short lived, but which some patients pretend is severely disabling because they 'enjoy the sick role'? What else could anyone think about an illness that cannot in any way be proved despite vast sums being spent on tests and that must be taken completely on faith. What else could anyone think about an illness that has seemingly been proven to be psychological or behavioural but where it seems patients would prefer to actually stay ill rather than to admit that they are mentally ill?

Every media article and government press release about 'CFS' is filled with fictional statements which make it very clear in many different ways that the illness has no scientific validity and that the patients do not deserve the same respect as other patient groups, but should be treated with contempt. Patients are not merely wrongly categorized as psychologically ill; it is so much more than that. It is persecution; patients are labelled as malingerers and deviants, and spoken about as if they were beneath contempt and not worthy of even basic respect or medical care, or even any level of kindness or compassion – even from their own friends and family. Whatever 'CFS' had been named, these problems would be the same. There is no such disease as 'CFS' and 'CFS' is merely an artificial entity created for the benefit of financial vested interest groups – that is the real problem, not the name 'CFS.'

What does the term ICD-CFS mean?

The various definitions of 'CFS' *do not* define M.E. Myalgic Encephalomyelitis as an organic neurological disorder as defined at G.93.3 in the World Health Organization's International Classification of Diseases (ICD). The definitions of 'CFS' do not reflect this. The 'CFS' or 'ME/CFS' definitions are not 'watered down' M.E. definitions, as some claim. They are not definitions of M.E. at all.

However, ever since an outbreak of M.E. in the US was given the label 'CFS,' the name/definition 'CFS' has prevailed for political reasons. 'CFS' is widely though wrongly applied to M.E. as well as to other diseases. The overwhelming majority of 'CFS' research does not involve M.E. patients and is not relevant *in any way* to M.E. patients. However, a minuscule percentage of research published under the name 'CFS' clearly does involve a significant number of M.E. patients as it details those abnormalities which are unique to M.E. Sometimes the problematic term 'ICD-CFS' is used in those studies and articles which, while they use the term 'CFS,' do relate to some extent to authentic M.E.

Problems with 'CFS' or so-called 'ICD-CFS' research

The overwhelming majority of 'CFS' research does not involve M.E. patients and is not relevant *in any way* to M.E. patients. A small number of 'CFS' studies refer in part to people with M.E. but it may not always be clear which parts refer to M.E. Unless studies are based on an exclusively M.E. patient group, results cannot be interpreted and are meaningless for M.E. While it is important to be aware of the small amount of research findings that do hold some value for M.E. patients, using the term 'ICD-CFS' to refer to this research is misleading and in many ways just damaging as using terms and concepts like 'ME/CFS' or 'CFS/ME.'

- For further details of the WHO ICD classifications of M.E. and 'CFS' worldwide and why terms such as 'ICD-CFS,' 'ME/CFS' and Myalgic 'Encephalopathy' must be avoided, please see the new paper by patient advocate Lesley Ben entitled: The World Health Organization's International Classification of Diseases (WHO ICD), ME, 'CFS,' 'ME/CFS' and 'ICD-CFS'
- Virtually all of the research which does relate to M.E., at least in part, but which uses the term/concept of 'CFS,' or ME/CFS, or CFIDS etc.— is also contaminated in some way by 'CFS' misinformation. Most often these papers contain a bizarre mix of facts relating to both M.E. and 'CFS.' For more information on some of the most common inaccuracies and 'CFS' propaganda included in this research, see the paper: Putting Research and Articles on M.E. into context and A warning on 'CFS' and 'ME/CFS' research and advocacy

What does define M.E.? What is its symptomatology?

M.E. is a systemic acutely acquired illness, initiated by a virus infection, which is characterised by post encephalitic damage to the brain stem (CNS) — a nerve centre through which many spinal nerve tracts connect with higher centres in the brain in order to control all vital bodily functions. This is always damaged in M.E., hence the name Myalgic Encephalomyelitis.

The CNS is diffusely injured at several levels; these include the cortex, the limbic system, the basal ganglia, the hypothalamus as well as areas of the spinal cord and its appendages. This persisting multilevel CNS dysfunction is undoubtedly both the chief cause of disability in M.E. and the most critical in the definition of the entire disease process.

M.E. represents an acute change in the balance of neuropeptide messengers, and consequently, a resulting loss of the ability of the CNS to adequately receive, interpret, store and recover information which enables it to control vital body functions (cognitive, hormonal, cardiovascular, autonomic and sensory nerve communication, digestive, visual auditory balance etc). It is a loss of normal internal homeostasis. The individual can no longer function systemically within normal limits.

M.E. is primarily neurological, but because the brain controls all vital bodily functions, virtually every bodily system can be affected by M.E. Again, although M.E. is primarily neurological it is also known that the vascular and cardiac dysfunctions seen in M.E. are the cause of many of the symptoms and much of the disability associated with M.E., and that the well-documented mitochondrial abnormalities present in M.E. significantly contribute to both of these pathologies. There is also multi-system involvement of cardiac and skeletal muscle, liver, lymphoid and endocrine organs in M.E. Some individuals also have damage to skeletal and heart muscle.

M.E. symptoms are manifested by virtually all bodily systems including: cognitive, cardiac, cardiovascular, immunological, endocrinological, respiratory, hormonal, gastrointestinal and musculo-skeletal dysfunctions and damage.

M.E. is an infectious neurological disease and represents a major attack on the CNS – and an associated injury of the immune system – by the chronic effects of a viral infection. There is also transient and/or permanent damage to many other organs and bodily systems in M.E.

M.E. affects the body systemically. Even minor levels of physical and cognitive activity, sensory input and orthostatic stress beyond an M.E. patient's individual post-illness limits causes a worsening of the illness, and of symptoms, which can persist for days, weeks, months or even longer. In addition to the risk of relapse, repeated or severe overexertion can also cause permanent damage (e.g. to the heart), disease progression and/or death in M.E.

M.E. is not stable from one hour, day, week or month to the next. It is the combination of the chronicity, the dysfunctions, the instability and the lack of dependability of these functions that creates the high level of disability in M.E. It is also worth noting that of the CNS dysfunctions, cognitive dysfunction is a major disabling characteristic of M.E.

All of this is not simply theory, but is based upon an enormous body of mutually supportive clinical information. These are well-documented, scientifically sound explanations for why patients are bedridden, profoundly intellectually impaired, unable to maintain an upright posture and so on (Chabursky et al. 1992 p. 20) (Hyde 2007, [Online]) (Hyde 2006, [Online]) (Hyde 2003, [Online]) (Hyde 2009) (Dowsett 2001a, [Online]) (Dowsett 2000, [Online]) (Dowsett 1999a, [Online]) (Hyde 1992 pp. x-xxi) (Hyde & Jain 1992 pp. 38 - 43) (Hyde et al. 1992, pp. 25-37) (Dowsett et al. 1990, pp. 285-291) (Ramsay 1986, [Online]) (Dowsett & Ramsay n.d., pp. 81-84) (Richardson n.d., pp. 85-92).

- *What is homeostasis?* Homeostasis is the property of a living organism, to regulate its internal environment to maintain a stable, constant condition, by means of multiple dynamic equilibrium adjustments, controlled by interrelated regulation mechanisms. Homeostasis is one of the fundamental characteristics of living things. It is the maintenance of the internal environment within tolerable limits.

What are some of the symptoms of M.E.?

More than 64 distinct symptoms have been authentically documented in M.E. At first glance it may seem that every symptom possible is mentioned, but although people with M.E. have a lot of different minor symptoms because of the way the central nervous system (which controls virtually every bodily system) is affected, the major symptoms of M.E. really are quite distinct and almost identical from one patient to the next (Hooper & Montague 2001a, [Online]) (Hyde 2006, [Online]). Individual symptoms of M.E. include:

Sore throat, chills, sweats, low body temperature, low grade fever, lymphadenopathy, muscle weakness (or paralysis), muscle pain, muscle twitches or spasms, gelling of the joints, hypoglycaemia, hair loss, nausea, vomiting, vertigo, chest pain, cardiac arrhythmia, resting tachycardia, orthostatic tachycardia, orthostatic fainting or faintness, circulatory problems, opthalmoplegia, eye pain, photophobia, blurred vision, wavy visual field, and other visual and neurological disturbances, hyperacusis, tinnitus, alcohol intolerance, gastrointestinal and digestive disturbances, allergies and sensitivities to many previously well-tolerated foods, drug sensitivities, stroke-like episodes, nystagmus, difficulty swallowing, weight changes, paresthesias, polyneuropathy, proprioception difficulties, myoclonus, temporal lobe and other types of seizures, an inability to maintain consciousness for more than short periods at a time, confusion, disorientation, spatial disorientation, disequilibrium, breathing difficulties, emotional lability, sleep disorders; sleep paralysis, fragmented sleep, difficulty initiating sleep, lack of deep-stage sleep and/or a disrupted circadian rhythm.

Neurocognitive dysfunction may include cognitive, motor and perceptual disturbances. Cognitive dysfunction may be pronounced and may include: difficulty or an inability to speak (or understand speech), difficulty or an inability to read or write or to do basic mathematics, difficulty with simultaneous processing, poor concentration, difficulty with sequencing, and problems with memory including difficulty making new memories, difficulty recalling formed memories and difficulties with visual and verbal recall (e.g. facial agnosia). There is often a marked loss in verbal and performance intelligence quotient (IQ) in M.E. (Bassett 2010, [Online]).

- For a more complete symptom list see: The ultra-comprehensive M.E. symptom list. See also: What it feels like to have M.E.: A personal M.E. symptom list and description of M.E.
- See the Research and Articles section for many hundreds of different articles and medical studies into M.E.

What other features define or characterise M.E.?

What characterises M.E. every bit as much as the individual neurological, cognitive, cardiac, cardiovascular, immunological, endocrinological, respiratory, hormonal, muscular, gastrointestinal and other symptoms is the way in which people with M.E. respond to physical and cognitive activity, sensory input and orthostatic stress;. -n other words, the pattern of symptom exacerbations, relapses and disease progression.

The way the bodies of people with M.E. react to these activities/stimuli post-illness is unique in a number of ways. Along with a specific type of damage to the CNS, this characteristic is one of the defining features of the illness and must be present for a correct diagnosis of M.E. to be made. The main characteristics of the pattern of symptom exacerbations, relapses and disease progression in M.E. include the following:

A. People with M.E. are unable to maintain their pre-illness activity levels. This is an acute, sudden change. M.E. patients can only achieve 50% or less of their pre-illness activity levels.

B. People with M.E. are limited in how physically active they can be but are also limited in similar ways with cognitive exertion, sensory input and orthostatic stress.

C. When a person with M.E. is active beyond their individual physical, cognitive, sensory or orthostatic limits, there is a worsening of various neurological, cognitive, cardiac, cardiovascular, immunological, endocrinological, respiratory, hormonal, muscular, gastrointestinal and other symptoms.

D. The level of physical activity, cognitive exertion, sensory input or orthostatic stress that is needed to cause a significant or severe worsening of symptoms varies from patient to patient, but is often trivial compared to a patient's pre-illness tolerances and abilities.

E. The severity of M.E. waxes and wanes throughout the hour/day/week and month.

F. The worsening of the illness caused by overexertion often does not peak until 24 - 72 hours or more later.

G. The effects of overexertion can accumulate over longer periods of time and lead to disease progression or death.

H. The activity limits of M.E. are not short term: an increase in activity levels beyond a patient's individual limits, even if gradual, causes relapse, disease progression or death.

I. The symptoms of M.E. do not resolve with rest. The symptoms and disability of M.E. are not caused only by overexertion: there is also a base level of illness which can be quite severe even at rest.

J. Repeated overexertion can harm the patient's chances for future improvement in M.E. Patients who are able to avoid overexertion have repeatedly been shown to have the most positive long-term prognosis.

K. Not every M.E. sufferer has 'safe' activity limits within which they will not exacerbate their illness: this is not the case for very severely affected patients.

- For the full-length version of this text and for a full list of references for this text see: The ultra-comprehensive M.E. symptom list.

What causes M.E.?

M.E. expert Dr Byron Hyde explains that:

> [The] prodromal phase is associated with a short onset or triggering illness. This onset illness usually takes the form of either, or any combination, of the following, (a) an upper respiratory illness, (b) a gastrointestinal upset, (c) vertigo and (d) a moderate to severe meningitic type headache. The usual incubation period of the triggering illness is 4-7 days. The second and third phases of the illness are usually always different in nature from the onset illness and usually become apparent within 1-4 weeks after the onset of the infectious triggering illness (1998 [Online]).

Despite popular opinion, (and the vast amount of 'CFS' government and media propaganda which purports to be relevant to M.E. but is not), there is **no** link between contracting M.E. and being a 'perfectionist' or having a 'type A' or over-achieving personality. M.E. **cannot** be caused by a period of long-term or intense stress, trauma or abuse in childhood, becoming run-down, working too hard or not eating healthily. M.E. is not a form of 'burnout' or nervous exhaustion, or the natural result of a body no longer able to cope with long-term stress.

Research also shows that it is simply not possible that M.E. could be caused by the Epstein-Barr virus, any of the herpes viruses (including HHV6), glandular fever/mononucleosis, Cytomegalovirus (CMV), Ross River virus, Q fever, hepatitis, chicken pox, influenza or any of the bacteria which can result in Lyme disease (or other tick-borne bacterial infections). M.E. is also not a form of chemical poisoning.

M.E. is undoubtedly caused by a virus, a virus with an incubation period of 4-7 days. There is also ample evidence that M.E. is caused by the same type of virus that causes Polio: an enterovirus (Hyde 2006, [Online]) (Hyde 2007, [Online]) (Hooper 2006, [Online]) (Hooper & Marshall 2005a, [Online]) (Hyde 2003a, [Online]) (Dowsett 2001a, [Online]) (Hooper et al. 2001, [Online]) (Dowsett 2000, [Online]) (Dowsett 1999a, 1999b, [Online]) (Ryll 1994, [Online]).

- See The outbreaks (and infectious nature) of M.E. section for more information.
- For information on the outrageous hype surrounding the recent XMRV 'CFS' research, please see the XMRV, 'CFS,' and M.E. paper by Sarah Shenk as well as the HFME press release: International M.E. expert disputes that 'CFS' XMRV retrovirus claim has relevance to M.E. patients

Are there outbreaks of M.E.?

One of the most fundamental facts about M.E. throughout its history is that it occurs in epidemics. There is a history of over sixty recorded outbreaks of the illness going back to 1934 when an epidemic of what seemed at first to be Poliomyelitis was reported in Los Angeles. As with many of the other M.E. outbreaks, the Los Angeles outbreak occurred during a local Polio epidemic.

The presenting illness resembled Polio, so for some years the illness was considered to be a variant of Polio and classified as 'Atypical Poliomyelitis' or 'Non-paralytic Polio' (TCJRME 2007, [Online]) (Hyde 1998, [Online]) (Hyde 2006, [Online]). Many early outbreaks of M.E. were also individually named for their locations so we also have outbreaks known as Tapanui flu in New Zealand, Akureyri or Icelandic disease in Iceland, Royal Free Disease in the UK, and so on (TCJRME 2007, [Online]) (Hyde 1998, [Online]).

A review of early M.E. outbreaks found that clinical symptoms were consistent in over sixty recorded epidemics spread all over the world (Hyde 1998, [Online]). Despite the different names being used, these were repeated outbreaks of the same illness. It was also confirmed that the epidemic cases of M.E. and the sporadic cases of M.E. each represented the same illness (Hyde 2006, [Online]) (Dowsett 1999a, [Online]).

M.E. is an infectious neurological disease and represents a major attack on the CNS by the chronic effects of a viral infection. The world's leading M.E. experts, namely Ramsay, Richardson, Dowsett and Hyde, (and others) have all indicated that M.E. is caused by an enterovirus.

The evidence which exists to support the concept of M.E. as an enteroviral disease is compelling (Hyde 2007, [Online]) (Hyde 2006, [Online]). An enterovirus explains the age variation, sex variation, obvious resistance of some family members to the infection and the effect of physical activity — particularly in the early stages of the illness — in creating more long-term/severe M.E. illness in the host (Hyde & Jain 1992a, p. 40).

There is also the evidence that:

- M.E. epidemics very often followed Polio epidemics.
- M.E. resembles Polio at onset.
- Serological studies have shown that communities affected by an outbreak of M.E. were effectively blocked (or immune) from the effects of a subsequent Polio outbreak.
- Evidence of enteroviral infection has been found in the brain tissue of M.E. patients at autopsy (Hyde 2007, [Online]) (Hyde 2006, [Online]) (Hyde 2003, [Online]) (Dowsett 2001a, [Online]) (Dowsett 2000, [Online]) (Dowsett 1999a, 1999b, [Online]) (Hyde 1992 p. xi) (Hyde & Jain 1992 pp. 38 - 43) (Hyde et al. 1992, pp. 25-37) (Dowsett et al. 1990, pp. 285-291) (Ramsay 1986, [Online]).

The US Centres for Disease Control (CDC) placed 'CFS' on its "Priority One, New and Emerging" list of infectious diseases some years ago; a list that also includes Lyme disease, hepatitis C, and malaria' (Gellman & Verillo 1997, p. 19). Despite this, no real research into transmissibility (or more importantly on reducing infection rates) has been done by any government on patients with M.E. (or even 'CFS') despite ample evidence that this is an infectious disease.

There have been many well-documented clusters or outbreaks of the illness, reports of as many as 4.5% of M.E. sufferers contracting the illness immediately after blood transfusions (or after needle-stick injuries involving the blood of M.E. patients) and evidence of the disease spreading through casual contact amongst family members(Johnson, 1996) (Carruthers et al. 2003, p.79).

As Dr Elizabeth Dowsett explains: 'The problem we face is that, in spite of overwhelming epidemiological and technical evidence of an infectious cause, the truth is being suppressed by the government and the 'official' M.E. charities as 'too scary' for the general public' (n.d.a, [Online]).

This pretence of ignorance on behalf of government worldwide has had enormous consequences: for example, only in the UK are people with M.E. specifically banned from donating blood. Consequently, the number of people infected with M.E. continues to rise unabated and largely unnoticed by the public.

- See: The outbreaks (and infectious nature) of M.E. page for more information.

Is M.E. difficult to diagnose? What tests can be used to diagnose M.E.?

M.E. is a distinct, recognisable disease entity that is not difficult to diagnose and can in fact be diagnosed relatively early in the course of the disease (within just a few weeks), providing that the physician has some experience with the illness. There is just no other illness that has all the major features of M.E.

Although there is as yet no single test which can be used to diagnose M.E., there are (as with Lupus, Multiple Sclerosis, ovarian cancer and many other illnesses) a *series* of tests which can confirm a suspected M.E. diagnosis. Virtually every M.E. patient will also have various abnormalities visible on physical exam. If all tests are normal, if specific abnormalities are not seen on certain of these tests (e.g. brain scans), then a diagnosis of M.E. cannot be correct (Hyde 2007, [Online]) (Hyde 2006, [Online]) (Hooper et al. 2001, [Online]) (Chabursky et al. 1992, p.22).

As M.E. expert Dr Byron Hyde explains:

The one essential characteristic of M.E. is acquired CNS dysfunction. A patient with M.E. is a patient whose primary disease is CNS change, and this is measurable. We have excellent tools for measuring these physiological and neuropsychological changes: SPECT, xenon SPECT, PET, and neuropsychological testing (2003, [Online]).

Tests which together can be used to confirm an M.E. diagnosis include:

- SPECT and xenon SPECT scans of the brain
- MRI and PET scans of the brain
- Neurological examination
- Neuropsychological testing (including QEEG scans)
- The Romberg or tandem Romberg test

- Various tests of the immune system (including tests of natural killer cells number and function)
- Insulin levels and glucose tolerance tests
- Sedimentation rate testing (M.E. is one of less than half a dozen diseases which can cause sedimentation rates as low as zero)
- Circulating blood volume tests (which may show a reduced circulating blood volume of up to 50%)
- 24 hour Holter monitor testing (a type of heart monitor)
- Tilt table examination and blood pressure tests
- Exercise testing and chemical stress tests
- Physical exam

These tests are the most critical in the diagnosis of M.E., although various other types of tests are also useful.

Dr Byron Hyde's highly regarded (and TESTABLE) M.E. definition The Nightingale Definition of M.E. makes diagnosis easier than ever before, even for those with no experience with the illness (Hyde 2007, [Online]) (Hyde 2006, [Online]) (Hooper & Marshall 2005a, [Online]) (Hyde 2003, [Online]) (Dowsett 2001a, [Online]) (Dowsett 2000, [Online]) (Hyde 1992 p. xi) (Hyde & Jain 1992 pp. 38 - 43) (Hyde et al. 1992, pp. 25-37) (Dowsett et al. 1990, pp. 285-291) (Ramsay 1986, [Online]) (Dowsett n.d., [Online]) (Dowsett & Ramsay n.d., pp. 81-84) (Richardson n.d., pp. 85-92).

- Objective scientific tests *are* available which can aid in the diagnosis of M.E. and easily prove the severe abnormalities across many different bodily systems seen in M.E. Unfortunately many patients are not given access to these tests. Problems also exist with doctors not being familiar with the abnormalities on testing seen in M.E. and so *misinterpreting* the results of some tests. The problem is not that these tests don't exist, but that doctors – and many patients – are unaware of this information on testing, that it is not generally accepted due to the nefarious influence of political and financial vested interest groups, and that there are overwhelming financial and political incentives for researchers to IGNORE this evidence in favour of the bogus 'CFS' (or 'subgroups of 'ME/CFS') construct. For more information see: Testing for M.E. and Are we just 'marking time?'

How common is M.E.? Who gets M.E. and how?

Although the illness we now know as M.E. has existed for centuries, for much of that time it was a relatively uncommon disease. Following the mass Polio vaccination programs of the 1960s, cases of Polio were greatly reduced and outbreaks of M.E. seemed to be similarly affected. It wasn't until the late 1970s that M.E. began its dramatic increase in incidence worldwide. Over 20 years later, M.E. is a worldwide epidemic of devastating proportions. Many people have died from M.E. and there are now many hundreds of thousands of people severely disabled by this epidemic (TCJRME 2007, [Online]) (Hyde 1992, p. xi).

The main period of infectivity of M.E. peaks at the time just before symptoms appear through to the initial acute phase of the illness (which lasts for several months or in some cases years). M.E. appears to be highly infective but also highly selective. The major mode of infectivity is by an airborne or respiratory route. Modes of transmission are thought to include: casual contact (respiratory), salivary transmission (e.g. kissing), sexual transmission and transmission through blood products (Hyde et al. 1992, pp. 25 - 37). (A recent study of 752 patients found that 4.5% of them – almost one in twenty – had had a blood transfusion days or a week before experiencing acute onset of M.E.) (Carruthers et al. 2003, [Online]) (Hyde et al. 1992, pp. 25 - 37).

M.E. has a similar strike rate (or possibly somewhat higher), to Multiple Sclerosis and is estimated to affect roughly 0.2% of the population. Children and teenagers are also susceptible to the illness and children as young as five have been diagnosed with M.E. (M.E. can occur in children younger than five, but this is thought to be rare.) All ages are affected but most commonly sufferers are under 45 at onset. Women are affected around three times as often as men, a ratio common in autoimmune disorders, although in children the sexes seem to be afflicted equally.

M.E. affects all ethnic and socio-economic groups and has been diagnosed all over the world. There are more than a million M.E. sufferers worldwide (Hooper et al. 2001 [Online]) (Hyde 1992, pp. x - xxi).

- The CDC has recently released vastly inflated estimates for figures affected by 'CFS' but it should be noted that the number of people suffering with sustained fatigue has no more relevance to patients with M.E. than to those with M.S. or AIDS or any other distinct illness. See: More medical 'firsts' from the CDC?

Are there any treatments for M.E.?

There are no easy or quick cures for M.E., nor are any on the horizon – despite a lot of hype about various fairly unpromising 'CFS' research endeavours. Intelligent nutritional, pharmaceutical and other interventions can make a significant difference to a patient's life, however.

Appropriate biomedical diagnostic testing should be done as a matter of course (and repeated regularly) to ensure that the aspects of the illness which are able to be treated *can* be diagnosed, monitored and then treated as appropriate. Testing is also important so that dangerous deficiencies and dysfunctions, which may place the patient at significant risk, are not overlooked (Hooper at al. 2001 [Online]). For specific information on M.E. treatment, the following HFME papers are recommended reading:

- Treating M.E. - The basics and Treating and living with M.E.: Overview
- Finding a good doctor when you have M.E.
- Symptom-based management vs. deep healing in M.E.
- A quick start guide to treating and improving M.E. with aggressive rest therapy, diet, toxic chemical avoidance, medications, supplements and vitamins
- Why research and try treatments when some groups claim an M.E. cure is coming soon?
- What if vitamin/mineral/protocol 'x' didn't work for me?
- Deep healing in M.E.: An order of attack!
- Treating M.E. in the early stages

What is known about M.E. so far?

There is an abundance of research which shows that M.E. is an organic illness which can have profound effects on many bodily systems. These are well-documented, scientifically sound explanations for why patients are bedridden, profoundly intellectually impaired, unable to maintain an upright posture and so on. More than **a thousand** good articles now support the basic premises of M.E. Autopsies have also confirmed such reports of bodily damage and infection (Hooper & Williams 2005a, [Online]).

Many different organic abnormalities have been found in M.E. patients (in peer reviewed research). Patient advocates Margaret Williams and Eileen Marshall explain that:

- There is evidence of disrupted biology at cell membrane level
- There is evidence of abnormal brain metabolism
- There is evidence of widespread cerebral hypoperfusion
- There is evidence of CNS and immune dysfunction
- There is evidence of CNS inflammation and demyelination
- There is evidence of hypomyelination
- There is evidence that M.E. is a complex, serious multi-system autoimmune disorder
- There is evidence of significant neutrophil apoptosis
- There is evidence that the immune system is chronically activated (e.g. the CD4:CD8 ratio may be grossly elevated)
- There is evidence that natural killer (NK) cell activity is impaired (i.e. diminished)
- There is evidence that the vascular biology is abnormal, with disrupted endothelial function
- There is novel evidence of significantly elevated levels of isoprostanes
- There is evidence of cardiac insufficiency and that patients are in a form of cardiac failure (which is exacerbated by even trivial levels of physical activity, cognitive activity and orthostatic stress)
- There is evidence of autonomic dysfunction (especially thermodysregulation; frequency of micturition with nocturia; labile blood pressure; pooling of blood in the lower limbs; reduced blood volume (with orthostatic tachycardia and orthostatic hypotension. Findings of a circulating blood volume of only 75% of expected are common, and in some patients the level is only 50% of expected.)
- There is evidence of respiratory dysfunction, with reduced lung function in all parameters tested
- There is evidence of neuroendocrine dysfunction (notably HPA axis dysfunction)
- There is evidence of recovery rates for oxygen saturation that are 60% lower than those in normal controls

- There is evidence of delayed recovery of muscles after exercise (affecting all muscles including the heart.)
- There is evidence of a sensitive marker of muscle inflammation
- There is evidence that the size of the adrenal glands is reduced by 50%, with reduced cortisol levels
- There is evidence of at least 35 abnormal genes, (these are acquired genetic changes, not hereditary), specifically those that are important in metabolism; there are more abnormal genes in M.E. than there are in cancer
- There is evidence of serious cognitive impairment (worse than occurs in AIDS dementia.)
- There is evidence of adverse reactions to medicinal drugs, especially those acting on the CNS.
- There is evidence that symptoms fluctuate markedly from day to day and even from hour to hour (2006, [Online])

Note that this is only a sample of some of the research available, not an exhaustive list.

It is known that M.E. is:

1. An acute onset (biphasic) epidemic or endemic infectious disease process
2. An autoimmune disease (with similarities to Lupus)
3. An infectious neurological disease, affecting adults and children
4. A disease which involves significant (and at times profound) cognitive impairment/dysfunction
5. A persistent viral infection (due to an enterovirus; the same type of virus which causes Poliomyelitis and post-Polio syndrome)
6. A diffuse and measurable injury to the vascular system of the CNS.
7. A CNS disease with similarities to M.S.
8. A variable (but always serious) diffuse, acquired brain injury
9. A systemic illness (associated with organ pathology; particularly cardiac)
10. A vascular disease
11. A cardiovascular disease
12. A type of cardiac insufficiency
13. A mitochondrial disease
14. A metabolic disorder
15. A musculo-skeletal disorder
16. A neuroendocrine disease
17. A seizure disorder
18. A sleep disorder
19. A gastrointestinal disorder
20. A respiratory disorder
21. An allergic disorder
22. A pain disorder
23. A life-altering disease
24. A chronic or lifelong disease associated with a high level of disability
25. An unstable disease: from one hour/day/week or month to the next
26. A potentially progressive or fatal disease

M.E. affects every cell in the body and almost every bodily system (Hyde 2007, [Online]) (Hooper et al. 2001, [Online]) (Cheney 2007, [video recording]) (Ramsay 1986, [Online]).

- For more information see the General articles and research overviews section. See also articles by: Dr Elizabeth Dowsett and Dr Byron Hyde.

Is there a legitimate scientific debate about whether or not M.E. is a 'real' neurological disease?

Despite popular opinion there simply is no legitimate scientifically motivated debate about whether or not M.E. is a 'real' neurological illness or not, or whether it has a biological basis.

The psychological or behavioural theories of M.E. and claims that M.E. is just another term for 'CFS' are no more scientifically viable than theories of a flat earth. They are pure fiction.

Are there any somewhat similar medical conditions?

There are a number of post-viral fatigue states or syndromes which may follow common infections such as mononucleosis/glandular fever, hepatitis, Q fever, Ross River virus and so on. M.E. is an entirely different condition to these self-limiting fatigue syndromes however, and it is *not* caused by the Epstein Barr virus or any of the herpes or hepatitis viruses. People suffering with any of these post-viral fatigue syndromes do not have M.E.

M.E. does have some limited similarities – to varying degrees – to illnesses such as Multiple Sclerosis, Lupus, post-Polio syndrome, Gulf War Syndrome and chronic Lyme disease, and others. But this does not mean that they represent the same etiological or pathobiological process. They do not.

M.E. is a distinct neurological illness with a distinct onset, symptoms, aetiology, pathology, response to treatment, long and short term prognosis, and World Health Organization classification (G.93.3) (Hyde 2006, [Online]) (Hyde 2007, [Online]) (Hooper 2006, [Online]) (Hooper & Marshall 2005a, [Online]) (Hyde 2003a, [Online]) (Dowsett 2001a, [Online]) (Hooper et al. 2001, [Online]) (Dowsett 2000, [Online]) (Dowsett 1999a, 1999b, [Online])

- See <u>M.E. and other illnesses</u> for more information. See also: <u>M.E. vs. M.S.: Similarities and differences.</u>

How well is research into M.E. research funded by government?

Governments around the world are currently spending $0 a year on M.E. research. Considering the severity of the illness and the vast numbers of patients involved, this is a worldwide disgrace.

- See <u>Putting research and articles on M.E. in context</u> and <u>A warning on 'CFS' and 'ME/CFS' research and advocacy</u> for more information about research into M.E. and the challenges involved. See the <u>Donations</u> page on the HFME website to make a donation towards M.E. research and advocacy.

Abuse and M.E.

Two of the most common interventions people with M.E. are encouraged to participate in are cognitive behavioural therapy (CBT) and graded exercise therapy (GET).

However, despite the misleading claims to the contrary made by various vested interest groups, no evidence exists which demonstrates that CBT and GET are appropriate, effective or safe treatments for M.E. patients. Studies by these groups (and others) involving miscellaneous psychiatric and non-psychiatric 'fatigue' sufferers, and their positive response to these treatments, have no more relevance to M.E. sufferers than they do to patients with Multiple Sclerosis, diabetes or any other illness. Patients with M.E. are routinely being prescribed these treatments on what amounts to a random basis medically.

As (very bad) luck would have it, graded exercise programs are probably the single most inappropriate 'treatment' that an M.E. sufferer could be encouraged to undertake. Permanent damage may result, as well as disease progression. Patient accounts of leaving exercise programs much more severely ill than when they began them are common: some end up wheelchair-bound, bed-bound or requiring hospitalisation in intensive care or cardiac care units. The damage caused is often severe and either long-term or permanent: some patients are still dealing with the effects of inappropriate advice to exercise five, ten or more YEARS afterwards, and for some patients this damage is permanent. Sudden deaths have also been reported in a small percentage of M.E. patients following exercise.

CBT and GET are at best useless and at worst extremely harmful for M.E. patients. Despite this, these 'treatments' are regularly recommended for people with M.E., who are assured that they are completely safe. Patient participation is not always voluntary. Many M.E. patients have been treated as psychiatric patients against their will (or against their parents' will in the case of children with M.E.). In some cases it is a condition of receiving medical insurance or government welfare entitlements that M.E. patients first undergo 'rehabilitation', including CBT and GET programs, particularly in the UK.

If a prescription drug had anything like the appalling track record exercise has with people with M.E. (or even a small fraction of it, even 2%) it would be a worldwide scandal. The drug would be immediately banned, there would be some form of inquiry and serious criminal charges may well be laid. Yet the rate of people with M.E. encouraged or even *forced* to exercise continues to rise, and with the full support of governments. This is despite the fact that legitimate research clearly shows that along with the huge risk involved, it has a zero percent chance of providing any benefit to people with authentic M.E. That this can be allowed to go on in such a supposedly enlightened day and age as ours defies belief.

It is also of great concern that so many M.E. patients are ONLY offered 'treatments' such as CBT and GET, while access to even basic appropriate medical care is withheld. Of the 30% of patients who are severely affected by the illness (and are bed-bound and housebound), the majority have no contact with the health service at all as they are seldom able to obtain house calls (Dunn 2005, [Online]). Many sufferers are also refused the basic welfare support to which they are entitled.

Thus a significant percentage of very physically ill and vulnerable M.E. patients are simply left to suffer and die at home without any medical care, welfare or social support (Hooper 2003a, [Online]).

- These brief comments on the effects of CBT and GET are taken from the more detailed paper: The effects of CBT and GET on patients with M.E., see this paper for more information.
- For more information about the effects of overexertion on M.E. patients, including statements/research from some of the world's leading M.E. experts about why overexertion is so physically harmful, see: Smoke and Mirrors. (This paper also includes links to M.E. patient accounts of the effects of overexertion).
- A recent example of an M.E. sufferer being taken into psychiatric care against their will is the case of Sophia Mirza, in the UK. Tragically, Sophia died of her illness after being wrongly sectioned under the Mental Health Act. Sophia was severely ill. and bedbound but she was refused even basic medical care, and this is believed to have contributed greatly to her death. For more information on this tragic case and entirely avoidable death, see: Inquest Implications, Civilization: Another word for barbarism by Gurli Bagnall and The Story of Sophia and M.E.
- For more information about forced exercise 'treatments' see the 100+ page CBT and GET Database. See also Comments on the 'Lightning Process' scam and other related scams aimed at M.E. patients

Is it only M.E. patients who are negatively affected by the bogus creation of 'CFS'?

If only. Vast numbers of patients from all sorts of varied patient groups misdiagnosed as' CFS' are also denied appropriate diagnosis and treatment, and may routinely be subjected to inappropriate psychological interventions such as CBT and GET. The 'CFS' insurance company scam also impacts negatively on doctors and the general public.

The only groups which gain from the 'CFS' confusion are insurance companies and various other organisations and corporations, including the government, which have a vested financial interest in how these patients are treated.

- For more information see: The misdiagnosis of 'CFS' and Who benefits from 'CFS' and 'ME/CFS'?

How severe is M.E.?

Although some people do have more moderate versions of the illness, symptoms are extremely severe for at least 30% of the people who have M.E., significant numbers of whom are housebound and bedbound.

Dr Paul Cheney stated before a US FDA Scientific Advisory Committee:

> I have evaluated over 2,500 cases. At worst, it is a nightmare of increasing disability with both physical and neurocognitive components. The worst cases have both an M.S.-like and an AIDS-like clinical appearance. We have lost five cases in the last six months. 80% of cases are unable to work or attend school. We admit regularly to hospital with an inability to care for self (Hooper et al. 2001 [Online]).

M.E. patients have been found to experience greater functional severity than the studied patients with heart disease, virtually all types of cancer, and all other chronic illnesses. In the 1980s Mark Loveless, an infectious disease specialist and head of the AIDS Clinic at Oregon Health Sciences University which also cared for patients with M.E., found that M.E. patients whom he saw had far lower scores on the Karnofsky performance scale than his HIV patients even in the last week of their life. He testified that an M.E. patient, 'feels effectively the same every day as an AIDS patient feels two weeks before death' (Hooper & Marshall 2005a, [Online]).

But in M.E., this extremely high level of illness and disability is not short-term. It does not always lead to death and it can instead continue uninterrupted for **decades**.

- For more information on severe M.E. see The severity of M.E. and M.E. fatalities and Why patients with severe M.E. are housebound and bedbound.
- Patients with M.E. may also find the following papers useful: Adjusting personal care tasks for the M.E. patient and The HFME M.E. ability and severity scale checklist
- If you would like a friend or family member to be included in the HFME M.E. memorial list, please see the HFME memorial lists page for contact details, and for further information.
- It should also be noted that even those patients with moderate M.E. are far more affected than many patients with a variety of other illnesses. Of course severe M.E. is even worse, but moderate M.E. can also cause significant symptoms and a relatively higher level of disability and suffering than many other illnesses.

Recovery from M.E.

M.E. patients who are given advice to rest in the early stages of the illness, and who avoid overexertion thereafter, have repeatedly been shown to have the most positive long-term prognosis.

As M.E. expert Dr Melvin Ramsay explains:

> The degree of physical incapacity varies greatly, but the [level of severity] is directly related to the length of time the patient persists in physical effort after its onset; put in another way, **those patients who are given a period of enforced rest from the onset have the best prognosis**. Since the limitations which the disease imposes vary considerably from case to case, the responsibility for determining these rests upon the patient. Once these are ascertained the patient is advised to fashion a pattern of living that comes well within them (1986, [Online]).

M.E. can be progressive, degenerative (change of tissue to a lower or less functioning form, as in heart failure), chronic, or relapsing and remitting. Some patients experience spontaneous remissions — albeit most often at a greatly reduced level of functioning compared to pre-illness — and such patients remain susceptible to relapses for the remainder of their lives. M.E. is a chronic/life-long disability where relapse is always possible. Cycles of severe relapse are common, as are further symptoms developing over time. Around 30% of cases are progressive and degenerative and sometimes M.E. is fatal. As Dr Elizabeth Dowsett writes:

> After a variable interval, a multi-system syndrome may develop, involving permanent damage to skeletal or cardiac muscle and to other "end organs" such as the liver, pancreas, endocrine glands and lymphoid tissues, signifying the further development of a lengthy chronic, mainly neurological condition with evidence of metabolic dysfunction in the brain stem. Yet, stabilisation, albeit at a low level, can still be achieved by appropriate management and support. The death rate of 10% occurs almost entirely from end-organ damage within this group (mainly from cardiac or pancreatic failure) (2001a, [Online]).

Clearly, many people with M.E. are significantly or severely disabled. But what is so tragic about this high level of suffering is that so much of it is needless. The appropriate support (financial, medical and practical) can do much to prevent the physical, occupational and deterioration in quality of life for M.E. patients and can stabilise the illness (Dowsett 2002b, [Online]).

Many deaths from M.E. could have been prevented if only those patients had been given a basic level of support and care made available to patients with illnesses with comparable care needs such as M.S. and Motor Neurone Disease.

- The 3 Part M.E. Ability and Severity Scale can be used to measure M.E. severity over time.
- For information on adrenaline surges in M.E., and the different order in which certain bodily systems may be affected by M.E. (and by overexertion), see the Dr Cheney section in The effects of CBT and GET on patients with M.E. or The importance of avoiding overexertion in M.E. (*Note that Dr Cheney does unfortunately mix M.E. and 'CFS' information and so cannot be considered an M.E. expert, as such.*)

Certain groups and individuals are benefiting enormously from this fraudulent artificial 'CFS' construct.

To say that these groups and individuals always believe what they are saying and that it is based on science or reality is ridiculous. To say that it is merely a misunderstanding or a mistake is equally ridiculous. The 'CFS' construct is a complete fiction, and exists purely because it is so financially and politically beneficial to a number of powerful groups.

The artificial 'CFS' construct is no more a scientifically accurate description of M.E. than it is a scientifically accurate description of M.S., Lupus or Polio. This pretence of ignorance about M.E. and about the reality of 'CFS', particularly by governments, has had devastating consequences for people with M.E. – as well as all of those with non-M.E. illnesses who are misdiagnosed as having 'CFS' – and has also meant that the number of M.E. sufferers continues to rise unabated and largely unrecognised. The general public worldwide, including sufferers themselves, has been lied to repeatedly about the reality of M.E.

The continuing, decades old, systemic abuse and neglect of the million or more people with M.E. worldwide has to stop. M.E. and' CFS' are *not* the same. Concepts such as 'ME/CFS,' 'CFS/ME,' Myalgic 'Encephalopathy' and 'CFIDS' are also unhelpful, unscientific and only add to the obfuscation.

'CFS' is merely a scam invented by insurance companies motivated by profit without regard for truth or ethics. These groups are acting without any regard for the extreme suffering and avoidable deaths they are causing. These groups are acting criminally. The scam is tissue thin and very easily discovered if one merely takes the time to look at the evidence.

Why is almost nobody doing this? Why is the world letting these groups get away with such a heinous scam and such appalling abuse on a massive scale? Why isn't the world caring enough or smart enough or gutsy enough to see through these slick, well-funded misinformation campaigns, and to act? How can this be, when the lies are so flimsy and scientifically laughable? Have we learned nothing from the devastating corporate cover-ups of the truth about tobacco and asbestos in our recent past? Where is the World Health Organisation? Where are our human rights groups? Where is our media? Where are our uncompromising investigative journalists?

Will it take another 20 years? How much more extreme do the suffering and abuse have to be? How many more hundreds of thousands of children and adults worldwide have to be affected? How many more patients will have to die needlessly before something is finally done? How much longer will we leave the fox in charge of the hen house? It's insupportable.

Where do we go from here?

Sub-grouping different types of 'CFS,' refining the bogus 'CFS' definitions further or renaming 'CFS' with some variation on the term M.E. would achieve nothing and create yet more confusion and mistreatment. The problem is not that 'CFS' patients are being mistreated as psychiatric patients; some of those patients misdiagnosed with' CFS' actually *do* have psychological illnesses.

There is no such distinct disease as 'CFS' – that is the entire issue, and the vast majority of patients misdiagnosed with' CFS' *do not* have M.E. and so have no more right to that term than to 'cancer' or 'diabetes.' The only way forward, for the benefit of society and every patient group involved, is that:

1. The bogus disease category of 'CFS' must be abandoned completely. Patients with fatigue (and other symptoms) caused by a variety of different illnesses need to be diagnosed correctly with these illnesses if they are to have any chance of recovery, and not given a meaningless 'CFS' misdiagnosis. Patients with M.E. need this same opportunity. Each of the patient groups involved must be correctly diagnosed and treated as appropriate, based on legitimate and unbiased scientific evidence involving the SAME patient group.

2. The name Myalgic Encephalomyelitis must be fully restored (to the exclusion of all others) and the World Health Organization classification of M.E. (as a distinct neurological disease) must be accepted and adhered

to in all official documentations and government policy. As Professor Malcolm Hooper explains:

> The term Myalgic Encephalomyelitis was first coined by Ramsay and Richardson and has been included by the World Health Organisation (WHO in their International Classification of Diseases (ICD), since 1969. The current version ICD-10 lists M.E. under G.93.3 - neurological conditions. It cannot be emphasised too strongly that this recognition emerged from meticulous clinical observation and examination (2006, [Online]).

3. People with M.E. must immediately stop being treated as if they are mentally ill or suffer with a behavioural illness; as if their physical symptoms do not exist or can be improved with 'positive thinking' and exercise, or be mixed in with various 'fatigue' sufferers or patients with any other illness than authentic Myalgic Encephalomyelitis. People with M.E. must also be given access to basic medical care, financial support and other appropriate services (including funding for legitimate M.E. research) on an equal level to that which is available for those with comparable illnesses (e.g. M.S. or Lupus). The facts about M.E. must be taught to medical students, and included in mainstream medical journals.

- See On the Name Myalgic Encephalomyelitis for more information on the evidence for inflammation of the brain and spinal cord in M.E. and other issues surrounding the name Myalgic Encephalomyelitis.
- See also Who benefits from 'CFS' and 'ME/CFS'?, Problems with the so-called "Fair name" campaign and Problems with the use of 'ME/CFS' by M.E. advocates, plus Problems with 'our' M.E. (or 'CFS' 'CFIDS' or 'ME/CFS' etc.) advocacy groups (also available in an animated video format.)

What can you do to help?

Unlike people with HIV/AIDS, people with M.E. do not have an initial period of their illness where they are only mildly affected. M.E. is severely disabling even in the first week of illness. People with M.E. are almost all far too ill to stage protests, rallies or marches. Many with M.E. cannot even read enough to be able to understand what is happening, and are not even aware that high quality scientific information on M.E. exists and that supporting the various 'CFS' and 'ME/CFS' faux 'advocacy' groups is counter-productive in the extreme.

Almost all so-called patient advocacy groups worldwide have sold patients out to the highest bidder and are now *actively collaborating* with our abusers. These groups are no longer advocates for patients with M.E. – indeed they are working directly AGAINST the interest of people with M.E. These groups also do not help all those misdiagnosed with 'CFS', who do not have M.E. The media too has sold-out and betrayed M.E. patients. People with M.E. have only a tiny minority of the medical, scientific, legal and other potentially supporting professions, as well as the public, on their side.

The Committee for Justice and Recognition of Myalgic Encephalomyelitis explains:

> There is no immunity to M.E. The next victim of this horrible disease could be your sister, your friend, your brother, your grandchildren, your neighbour [or] your co-worker. M.E. is an infectious disease that has become a widespread epidemic that is not going away. We must join together, alert the public and demand action (2007, [Online]).

That is what is needed – people power. Educated people power. For people from all over the world to stand up for M.E. Individual physicians, journalists, politicians, human rights campaigners, patients, families and friends of patients and the public, whether they are affected yet by M.E. or not, must stand up for the truth. That is the only way change will occur— through education and people simply refusing to accept what is happening any more.

Yes, there are powerful and immensely wealthy vested interest groups out there, who will fight the truth every step of the way, but we have science, reality and ethics on our side and those are also very powerful. However, for this to be of any use to us, we must first make ourselves aware of the facts *and then use them.* **So what you can do to help is to PLEASE spread the truth about M.E. and try to expose the lie of 'CFS.'**

You can also help by NOT supporting the bogus concepts of 'CFS,' 'ME/CFS,' 'subgroups of ME/CFS,' 'CFS/ME,' 'CFIDS' and Myalgic 'Encephalopathy.' Do not give public or financial help or support to groups which promote these harmful and unscientific concepts or which equate M.E. with 'CFS.'

The abuse and neglect of so many seriously ill people on such an industrial scale is truly inhumane and has already gone on for far too long. People with M.E. desperately need your help.

References

All of the information concerning Myalgic Encephalomyelitis on this website is fully referenced and has been compiled using the highest quality resources available, produced by the world's leading M.E. experts. More experienced and more knowledgeable M.E. experts than these – Dr Byron Hyde and Dr Elizabeth Dowsett in particular – do not exist.

Between Dr Byron Hyde and Dr Elizabeth Dowsett, and their mentors the late Dr John Richardson and Dr Melvin Ramsay (respectively), these four doctors have been involved with M.E. research and M.E. patients for well over 100 years collectively, from the 1950s to the present day. Between them they have examined more than 15 000 individual (sporadic and epidemic) M.E. patients, as well as each authoring numerous studies and articles on M.E., and books (or chapters in books) about M.E. Again, more experienced, more knowledgeable and more credible M.E. experts than these simply do not exist.

This paper is intended to provide a brief summary of the most important facts of M.E. It has been created for the benefit of those people without the time, inclination or ability to read each of the far more detailed and lengthy references created by the world's leading M.E. experts. The original documents used to create this paper are essential additional reading, however, for any physician (or anyone else) with a real interest in Myalgic Encephalomyelitis. For a full reference list please see the References page of this book.

Acknowledgments
Thanks to Peter Bassett, Emma Searle and Lesley Ben for editing this paper.

Relevant quotes

'The problem with fatigue is that it is neither specific, definable nor scientifically measurable. Fatigue is both a normal and a pathological feature of every day life. Every normal person gets fatigued. Fatigue is a common feature of much major psychiatric disease and major medical disease. Since fatigue is such an integral part of many illnesses, by calling fatigue the primary characteristic, the authors necessitated the elimination of hundreds of other diseases. To truly follow the criteria set out by the CDC definition probably makes 'CFS' the most expensive illness to investigate of any known disease. Fatigue is not an object, it is simply a modifier in search of a noun. Also, taking fatigue as the flagship symptom of a disease not only bestows the disease with a certain Rip Van Winkle humour, but it removes the urgency of the fact that the majority of M.E. symptoms are in effect CNS symptoms. M.E. represents a major attack on the CNS by the chronic effects of a viral infection.'
DR BYRON HYDE IN 'THE CLINICAL AND SCIENTIFIC BASIS OF M.E. P 11-12

'Western newspapers and magazines are packed with trivia, television news is concealing the reality of what is happening… and investigative journalism has virtually died a death. [But] what is the point of democracy if you keep the citizens in a state of semi-ignorance?'
VETERAN ACTIVIST, PROTESTER AND AUTHOR TARIQ ALI

CHAPTER FIVE

Information about HFME, acknowledgments, references and final comments

This chapter includes the following papers and sections:

1. How, when and why HFME was founded

2. The aims of HFME and the reasons for the aims of HFME

3. Why hummingbirds as a metaphor for M.E.?

4. Personal acknowledgments

5. Acknowledgments

6. The HFME reference list

7. Additional HFME resources available online

8. Afterword

How, when and why HFME was founded

HFME was founded in May 2009. The leader and founder of the group is Jodi Bassett. Jodi Bassett is an Australian writer, artist, graphic designer, and patient advocate.

Jodi contracted Myalgic Encephalomyelitis (M.E.) in 1995 when she was just 19. She went from being healthy and happy one day, to very ill and disabled with the neurological disease M.E. the next.

When first ill, Jodi was reduced to 40% of her pre-illness activity level. Due to inappropriate medical advice leading to sustained overexertion (which causes serious and permanent bodily damage in M.E.), Jodi's illness quickly went from moderate to extremely severe. By 1999 she was capable of less than even 5% of her pre-illness activity level.

After more than a decade of the disease becoming worse as each year passed, Jodi's disability level finally began to stabilise. Thanks to appropriate care, education and support, her condition improved from *extremely severe* to *severe* in 2007. At the time of writing her condition continues, with careful management, to improve very slowly month by month.

She still requires the help of part-time carers to live, and is currently severely affected, housebound and largely bedbound. All of her activism and advocacy has been conducted from her bed using a laptop and a reclining laptop stand. Jodi is at best able to spend just 30 to 45 minutes a day (on average) on M.E. advocacy.

In 2004, Jodi Bassett started the 'A Hummingbirds' Guide to M.E.' website to try to improve awareness of the facts of M.E., and to stop other M.E. patients from being needlessly made far more ill and disabled due to inappropriate medical advice based on the false notion that M.E. is the same thing as 'CFS.'

In 2009, with the help of a group of similarly-minded M.E. advocates from around the world, Jodi founded 'The Hummingbirds' Foundation for M.E.' in order to advocate for M.E. patients on a much bigger scale and to get the relevant information to a much wider audience worldwide.

For the same or similar reasons, the majority of HFME contributors are likewise disabled.

There is very little advocacy for M.E. patients, and HFME contributors have determined that despite their high disability levels, they must do what they can for M.E. advocacy.

The vast majority of charities that started out advocating for M.E. patients are now actively supporting the same misinformation they were created to oppose. This is helped immeasurably by the bogus concept of 'ME/CFS.' For 20 years now, M.E. patients have been subjected to serious medical neglect and abuse, even unto death in some cases. The situation is actually worsening year by year as slick, faux advocacy groups gain more and more popularity and support from uneducated and misinformed – and often misdiagnosed – patients.

HFME is run by and for M.E. patients.

HFME contributors also aim to advocate for those non-M.E. patients given the always meaningless 'CFS' diagnosis who also are not being served well by the various 'CFS' charities, and who are also harmed by the bogus disease category of 'CFS' and the overwhelming triumph of financial greed over ethics, science and basic human rights.

In addition to Jodi Bassett, major HFME contributors include:

- Lesley Ben, a UK M.E. patient and patient advocate (the author of this paper)
- Ginny B, an Australian patient and patient advocate
- Emma Searle, an Australian M.E. patient and patient advocate
- Reverend Barbara Le Rossignol, LL.B.. B.D., an Australian M.E. patient and patient advocate
- Sarah Shenk, a US M.E. patient and patient advocate
- Vanessa Vaughn, an Australian M.E. patient and patient advocate

'The greater danger for most of us is not that our aim is too high and we miss it, but that it is too low and we reach it.'
MICHELANGELO BUONARROTI

The aims of HFME

 **Aim 1:** To disseminate scientifically accurate information on Myalgic Encephalomyelitis (M.E.) to M.E. patients; to their carers, family and friends; to the medical profession and other professions which deal with M.E. patients; to policy makers; to M.E. advocates and activists and to the general public, as per the paper <u>What is M.E.?</u> and as further discussed on the HFME website and in the HFME books.

Reason for Aim 1: An important fact about M.E. is that it can be made very much worse by overexertion. Overexertion can also cause death in M.E. It is vital that M.E. patients should be aware of the importance of avoiding overexertion in order to avoid needless permanent bodily damage. Many M.E. patients have become severely affected, bed-bound, or in constant pain because of overexertion and some patients have died. Tragically, these negative outcomes happen all too often, due to inappropriate medical advice.

M.E. patients currently receive little or no helpful medical advice, as most doctors have little understanding of the disease. Therefore it is essential that advice from M.E. experts on treatment and management of the disease should be made available to patients.

Aim 2: To oppose false and meaningless disease categories such as 'CFS,' 'CFIDS,' 'ME/CFS,' 'CFS/ME,' 'ME-CFS' and Myalgic 'Encephalopathy,' as per the papers <u>What is M.E.?</u> and <u>M.E. is not fatigue, or 'CFS'</u> and as further discussed on the HFME website and in the HFME books. These bogus disease categories and concepts must be abandoned for the benefit of all the different patient groups involved.

Reason for Aim 2: The fact that a person receives a diagnosis of 'CFS' (a) does not mean that the patient has M.E., and (b) does not mean that the patient has any other distinct and specific illness named 'CFS.' 'CFS' can only ever be a misdiagnosis, and prevents patients from getting a correct diagnosis and appropriate (even curative or life-saving) treatment. See: <u>The misdiagnosis of CFS</u>.

M.E. and 'CFS' are in no way synonymous terms. M.E. is a distinct, scientifically testable and measurable neurological disease which occurs in epidemic and sporadic forms. 'CFS' is a wastebasket diagnosis based on the presence of the symptom of fatigue. If serious abnormalities are found on testing, a person no longer qualifies for a 'CFS' diagnosis. For further information on the difference between M.E. and 'CFS,' see <u>What is M.E.?</u>

Currently few doctors or researchers recognise M.E. Patients receive little or no helpful medical advice, as most doctors have an entirely inaccurate understanding of the disease based on confusion with the bogus disease category of 'CFS.'

Concepts such as 'ME/CFS,' 'CFS/ME' and 'ME-CFS' are just as problematic and meaningless as 'CFS' and in many ways more so, as they incorrectly imply that M.E. and 'CFS' are synonymous terms. The only groups which gain from the continuation of these fictional disease categories (to the detriment of patients) are vested interest groups, such as:

1. Medical insurance companies
2. Governments
3. The vaccine industry
4. The chemical industry
5. Psychiatrists
6. 'CFS specialists'

7. Medical doctors
8. The media (including medical journals)
9. 'CFS' or 'ME/CFS' (and other) groups that sell vitamins and other supplements to 'CFS' patients
10. 'CFS' or 'ME/CFS' so-called patient support and advocacy groups

For more information see: Who benefits from 'CFS' and ME/CFS'?

Aim 3: To broaden the online and offline presence of HFME in order to disseminate information about M.E. and to correct misinformation about 'CFS' as per aims 1 and 2 above. This will involve improving internet accessibility as well as raising the profile of the website so that it can be found easily by M.E. patients as well as those misdiagnosed with 'CFS' who have other diseases. The information on the HFME website will also be made available in a convenient book format. Several book releases are planned.

Reason for Aim 3: There are currently very few websites or books available which accurately describe the historical, political and medical facts of M.E. and of 'CFS' and which also offer research papers by the world's leading M.E. experts, papers on various aspects of the disease (which are vital reading for patients and carers) and factual information and support. The HFME site and books aim to fill this gap.

For every factual website there are at least a hundred websites which support the misinformation about M.E. and 'CFS' which causes patients so much harm. The same is true about books on this subject.

M.E. patients suffer greatly from a lack of information on their condition, and urgently need the information offered by the HFME website and books, as do those patients misdiagnosed with 'CFS' who have other diseases. This information also urgently needs to become readily available to doctors, carers, human rights groups, the media and politicians.

Due to the restrictions commonly experienced by contributors and readers with M.E., HFME will primarily disseminate the facts of M.E. via the internet and in published books available for sale online and offline. However, HFME will also endeavour to participate in offline activism as much as possible (if it is possible at all, which is far from certain due to the very high disability levels of almost everyone involved currently) in order to reach a wider and more varied audience. This may include planning protests and rallies, participation in media interviews, distributing printed information on M.E. to the public, publically offering support to patients and their families facing persecution due to the involvement of vested interest groups with M.E., organising nationwide or worldwide M.E. events on set days, organising fundraising events, and so on.

Aim 4: To make it clear that M.E. is not 'medically unexplained' or 'mysterious' as 'CFS' is and that an abundance of scientific evidence already exists which proves that M.E. is a disabling and potentially fatal neurological disease. HFME makes available this valuable research which is generally overlooked.

Reason for Aim 4: The historical facts of M.E. and the available scientific research on M.E. is being **actively suppressed** and **deliberately ignored** due to vested political interests.

The scientific research on M.E. is generally overlooked in favour of misinformation promoted by vested interest groups incorporating fatigued patients who have been given a misdiagnosis of 'CFS.' Vested interest groups also disseminate outright propaganda and untruths for financial gain. For more information see: Who benefits from 'CFS' and ME/CFS'?

There is an abundance of scientific research on M.E. dating back to 1934, and much of what happens to the body with M.E. is understood, but since 1988, the distinct neurological disease M.E. has been confused with the fictional disease category of 'CFS'. The available science on M.E. is actively suppressed and ignored due to involvement by vested interest groups. Currently few doctors or researchers are aware of even the most basic facts about M.E.

Aim 5: To defend the M.E. community (and those with non-M.E. diseases misdiagnosed as 'CFS') against counter-productive 'activism' strategies such as renaming 'CFS' with some variation of the term M.E.

Reason for Aim 5: There are various 'activism' strategies based on false premises which promote harmful and unscientific terms and concepts such as 'ME/CFS' and Myalgic 'Encephalopathy.' Such terms and concepts obscure the reality of M.E., equate it with 'CFS' and harm the cause of M.E. These unscientific terms and concepts also cause harm to those with non-M.E. diseases misdiagnosed as 'CFS.'

Aim 6: To promote appropriate research based on a proper understanding of M.E., and to oppose flawed concepts such as the 'subgroups' of 'CFS' or 'ME/CFS' concept.

Reason for Aim 6: Research is unfortunately being attempted based on the false 'subgroups' of 'CFS' or 'ME/CFS' concept. M.E. is not a subgroup of 'CFS' or 'ME/CFS.' Studying subgroups of heterogeneous groups of fatigue patients does not in any way help M.E. patients, or any other distinct patient group.

Research based on flawed concepts and heterogeneous patient groups may also cause significant harm when it is wrongly thought to apply to distinct patient groups (as often occurs). This is particularly true with regard to M.E. M.E. patients must be treated based only on scientifically legitimate studies involving a 100% M.E. patient population.

Aim 7: To be a voice for those suffering from M.E. who are facing mistreatment and abuse due to the false notion that M.E. is the same thing as 'CFS' and is a trivial and 'mysterious' illness or a mental illness characterised by 'fatigue.'

Reason for Aim 7: Many patients with M.E. are subjected to medical abuse, mistreatment from social services, lack of understanding from the general public and even ridicule, neglect and abuse from friends and family. M.E. patients are (with very few exceptions) not being served by the charities which are supposed to represent them. The vast majority of charities purporting to help M.E. patients worldwide, whatever their original aims, are now actively supporting the propaganda which harms M.E. patients.

Aim 8: To be a voice for all those patients misdiagnosed with 'CFS' who do not have M.E., but other illnesses including: cancer, fibromyalgia, various post-viral fatigue syndromes, athlete's over-training syndrome, Lyme disease, Behcet's disease, PTSD, depression and other mental illnesses, burnout, thyroid or adrenal diseases, various vitamin-deficiency diseases, and so on. To encourage each of these patients to reject their 'CFS' misdiagnosis and seek a correct diagnosis and appropriate treatment.

Reason for Aim 8: A diagnosis of 'CFS' can only ever be a misdiagnosis. Currently, many hundreds of thousands of patients with a vast array of different diseases are misdiagnosed with 'CFS' and this can cause needless suffering and disability, sometimes leading to needless deaths.

There are many 'CFS' advocacy groups out there, but most tell patients that their 'CFS is real' and that 'CFS is not a mental illness' and that they must join the fight to legitimise 'CFS' and so on. This may placate patients in the short term, and make them feel as if they are being helped, but it misses the point entirely and is counter-productive. This approach keeps patients ignorant of the basic information that can improve their situation and their medical outcome.

Defending 'CFS' only aids vested interest groups; patients will benefit when 'CFS' is abandoned. Every 'CFS' misdiagnosed patient deserves to know that 'CFS' does not exist and that it should never be considered the end point of diagnosis.

Some of the conditions commonly misdiagnosed as 'CFS' are well defined and well-known illnesses that are treatable, but only once they have been correctly diagnosed. Some conditions are also very serious or can even be fatal if not correctly diagnosed and managed. Every patient deserves the best possible opportunity for appropriate treatment for their illness, and for recovery. This process must begin with a correct diagnosis. A correct diagnosis is half the battle won.

Aim 9: To enlist the help of human rights groups, medical professionals and the quality media to help to achieve the above stated goals as is their obligation and duty, a duty that has unfortunately been almost completely ignored for the last 20 years (with a few notable exceptions).

Aim 10: To obtain funding to pursue the aims described previously.

Quotes

'To the very few physicians still practicing today who began seeing patients with this illness some 40 years ago and who have continued to record and publish their clinical findings throughout, the current enthusiasm for renaming and reassigning this serious disability to subgroups of putative and vague "fatigue" entities, must appear more of a marketing exercise than a rational basis for essential international research. It was not always so unnecessarily complicated!'
REDEFINITIONS OF M.E. - A 20TH CENTURY PHENOMENON BY DR ELIZABETH DOWSETT

'Over the course of two International Association of Chronic Fatigue Syndrome (IACFS) conferences, there have been suggestions that the name CFS be changed to M.E., while retaining the CFS definitions as a basis for such change. This does not seem to me to be a useful initiative: it would simply add credence to the mistaken assumption that M.E. and CFS represent the same disease processes. They do not.'
DR BYRON HYDE 2006

'Advertising fatigue or studying it or analyzing it by fatigue scales or holding fatigue conferences or setting up fatigue clinics is not going to help those with underlying neurological illnesses.'
JILL MCLAUGHLIN

'Recently an M.E. patient's spine has been examined in the UK and the inflammatory nature was also discovered. Myalgic Encephalitis is a diffuse inflammatory injury of the capillaries at the level of the basement membrane of the brain. It makes no sense to rename the horse and call it Myalgic Encephalopathy. All brain pathologies involving brain tissue are encephalopathies. Let us stop fussing around and get back to the real problem and that is investigating the patients.'
DR BYRON HYDE

'Up to 1955, recognised M.E. was clearly previously associated with poliomyelitis. The viruses that cause paralytic poliomyelitis are some of the same viruses that cause M.E. But these enteroviruses that are capable of causing paralysis attach to more than one set of tissue receptors. These other receptors are found on different cells in the brain and spine as well as in other body areas. The symptoms described by M.E. sufferers are due to injury to these other cells.'
THE CLINICAL AND SCIENTIFIC BASIS OF M.E. DR BYRON HYDE P. 115

'Demand that the national health agencies end the policies and practices of deliberate roadblocks, distraction and dishonesty. These policies have had enormous consequences and continue to extract a tragic toll. Many people have died from Myalgic Encephalomyelitis and there are now millions disabled by this epidemic. It is time to put an end to the cover-up of this epidemic.'
THE COMMITTEE FOR JUSTICE AND RECOGNITION OF MYALGIC ENCEPHALOMYELITIS

'The world is a dangerous place to live; not because of the people who are evil, but because of the people who don't do anything about it.'
ALBERT EINSTEIN

'I must admit that I personally measure success in terms of the contributions an individual makes to her or his fellow human beings.'
MARGARET MEAD

'Never lose an opportunity of urging a practical beginning, however small, for it is wonderful how often in such matters the mustard-seed germinates and roots itself.'
FLORENCE NIGHTINGALE

Hummingbirds

Some time ago I was flicking through a book (looking for some artistic inspiration) when I came upon a stage-by-stage illustration of hummingbirds hovering and it struck a chord in me. Soon it hit me why. In the same way a hummingbird comes crashing to the ground with a big SPLAT! if it falters in the complex series of movements that keep it in the air, in a different sort of way, so do I.

I contracted Myalgic Encephalomyelitis (M.E.) in 1995 when I was 19. One day I was healthy and the next day absolutely everything changed. Since then I've been forced to keep on 'flapping my wings' endlessly lest I fall into an even more agony-filled and semi-conscious paralysed heap. I have to constantly remain aware of, and quickly adjust to, all sorts of small changes in my environment and my body. My version goes something like this;

FLAP! Making sure I don't spend too much time flat in bed (consecutively), or my vertigo becomes much more severe, the room spins horribly and I feel I am falling over backwards as I try to walk, or have to struggle not to fall off the edge of my perfectly flat bed.

FLAP! Trying not to stand or sit up for too long or my heart just can't cope and it struggles to beat properly and I feel extremely ill for hours afterward. It feels like a heart attack in every organ. Tests show my heart-rate can climb as high as 170 beats-per-minute just from a few minutes of 'exertion.'

Then I forget for just a few moments about having to be careful about how much light I expose my eyes to and instantly...C
      R
        A
          S
            H!

Burning pain that lasts for hours leaving me unable to open my eyes. But still I can't let my guard down and have to get myself back in the air straight away...

FLAP, FLAP, FLAP! I manage to quickly close all my doors and put my headphones on to block out some neighbourhood noise that would have left me in agonising pain, experiencing seizures, memory loss and taking five days to recover from if I'd listened to it at full volume.

Then I forget to avoid one of the foods I am intolerant of (but that I tolerated perfectly well the day before) and a few minutes later...**T H U D!** Abdominal pain, headache, bloating, severe itching and nausea for hours afterward. But quickly I have to get myself back up in the air...

FLAP! I manage to make my bath neither too cold (which leaves me shaking and unable to get warm for hours), or too hot (which makes me light-headed, my heartbeat becoming irregular and fluttery for the next six hours so it feels like my heart is struggling to beat and I'm having a heart attack.)

Then I forget to put my blanket over myself properly and within a short while...**S P L A T!** I get so cold I can't get myself warm again and it turns into a horrible shivering fever, which leads to delirium, paralysis and eventually loss of consciousness for several hours. I then spend the rest of the day partly paralysed and feeling (neurologically and cognitively) as if I'd had a stroke.

Because so many normal everyday things cause me to 'crash' I have to constantly monitor every little thing I do and every aspect of my environment to try and keep myself 'in the air' as much as possible. It's a never-ending task and a fairly thankless one too, as my highest level of functioning is pretty low anyway – I'm 100% house-bound and 99% bed-bound on my _best_ days, I'm in continual pain and experience many different neurological, cardiac, cognitive and other symptoms constantly. But it's not so much having a painful and limited life that is so hard to bear (though obviously that's part of it), but to have to plan and work so endlessly hard every minute just to keep my life this 'good': that's what _really_ makes it a nightmare.

With a bit more research however, I quickly found a more positive reason to identify with hummingbirds. You see, although at first glance they are tiny, seemingly defenceless and extremely vulnerable to attack from anyone or anything, they are actually quite tough little critters. They never back down from a fight even if the odds are against them, taking on other birds much larger than they are when they need to. What their bodies lack in strength and power is made up for by their bravery, strength of mind and _spirit_.

I've met so many people with M.E. that share that same spirit, particularly with severe M.E.; people that have remained kind, witty, giving, optimistic and determined to make the best of what they have despite dealing with an unbelievably severe (and potentially fatal) neurological, cardiac and metabolic disease often without the support of family, friends or the health and welfare systems; indeed, often with direct opposition, criticism and sometimes abuse from these people and organisations.

I consider these people no less beautiful inside than a hummingbird is to the eye. The human spirit is capable of amazing resilience and endurance and I can see no greater example of this than people suffering from severe M.E. I think they are truly inspiring. When every hour of every day is so difficult and there's no foreseeable end in sight, the fact that along with the obvious sadness and frustration there can also be hope and humour is just amazing.

The world is full of inspiring stories about people triumphing over quite small problems (compared to M.E.), always with the support of everyone around them and much back patting and praise when they've finished their short 'ordeal'. There is nothing wrong with that, except that on the other hand, there are desperately ill people with severe M.E. who have no support at all yet are able to somehow keep going through one horrendous ordeal of a day after another, often for many years or even decades. Not only are they rarely acknowledged for their hard work and amazing strength, but they are sometimes actually labelled as malingerers, or seen as mentally weak or defective in some way. Sometimes they are, unbelievably, treated as if they were merely very 'fatigued' or 'tired all the time' instead of very ill with a severe neurological and cardiovascular disease. It really does boggle the mind that there can be such a gap between perception and reality.

I've since featured hummingbirds in many of my paintings and drawings and this is why. I see the same sort of strength and beauty, combined with such heartbreaking vulnerability in my M.E. friends everyday. Nothing I've seen on this earth is more inspiring to me, more beautiful, _or_ more tragic, more heartbreaking. **I think of people with severe M.E. as hummingbirds now** – vulnerable, strong and strikingly beautiful all at once; and more than overdue for some consideration, compassion and care in this world.

This paper is dedicated to all the beautiful 'hummingbird' M.E. patients I've met. To view more of my artwork featuring hummingbirds please visit the www.ahummingbirdsguide.com website.

Acknowledgments
Thanks to Caroline Gilliford and Emma Searle for editing this paper.

Personal acknowledgments

While this book has a formal HFME acknowledgments section, I want to include some brief personal acknowledgments too. This book would not have been possible without the help and support of a lot of different people.

Thank you to my wonderful parents for all your support. Without your support the website would probably only be 5% of what it is, there would be no HFME books and my health would still be deteriorating, rather than very slowly improving. I wish everyone with M.E. could have the support, kindness and care you have given me. I am so grateful for all that you have done and continue to do for me, and for the people you are.

Thank you to my sister Claire Bassett (and her partner Dan) for being a better friend and sister than I could have ever hoped for.

A big thank you to my lovely friends Nicole Clarke, Vanessa Schneider, Cara Usher, Lesley Ben and Brooke Rodgers – and lately, to my brother Mark Bassett (and his partner Betty).

Thank you to Jo, Ginny, Emma, Els, Ingeborg, Clytie, Frír, Dianne, Bea, Sarah, Lottee, Vanessa, Stef, Charmion, Gurli, Simon, Anita, Jayne, Julie and everyone in the HFME/MEites United chat groups that has offered support, camaraderie and ideas over the years.

Thank you to all those hundreds, perhaps thousands, of M.E. patients and parents who have taken the time to discuss all aspects of M.E. with me, in so many private emails and online group discussions over the last 11 years. Thank you to everyone that has deciphered my badly-typed 'word-salad' emails!

Thank you to Daya and Joni for your friendship and all your kind, generous, beautiful and joyful mailings. I'm still in awe every time I get one in the post.

Thank you to my dear M.E. friends Sharon O'Day and Aylwin Catchpole, who both died in 2010. You were both 'friends in a million.' Thank you for the support you gave both HFME and me personally. It meant so much. You're both sadly missed, and will never be forgotten.

Thank you to my current doctor and to all those pioneers who have written books that have taught me so much and helped me improve my health, and give hope for further improvement in the future.

Thank you so much to Drs Byron Hyde and Elizabeth Dowsett. I can't imagine how difficult M.E. advocacy would be without your hard work. There would be no need for a group like HFME in 2011 if we'd had 100 of each of you all these years. I'm enormously grateful to you both personally, and to your mentors the late Dr Ramsay and Dr Richardson as well.

Finally, thank you to my furry feline friends that have made me laugh and kept me company for thousands of hours when I would otherwise have been completely alone. Thank you too to my extended family and to my old school and art school friends for your many kindnesses and considerations.

After writing my first paper on M.E. in 2004 I suffered excruciating pain, heart problems, seizures, blackouts and loss of vision for weeks and months. I never could have imagined back then being able to create a site like the HFME website is today, or writing this book. Something I began out of utter personal desperation is now something positive that I do out of choice, to try and help others.

I'd understand it if someone were to look at my contributions to HFME and be sceptical that they could have been done by someone so ill and disabled. It really is amazing what 30 minutes of work a day and lots of late-night musings, list-writing and notebook scribblings amount to after 7 solid years.

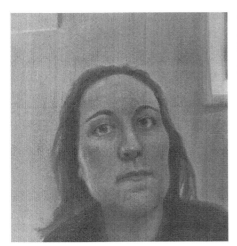

While I'd really love to be given access to my pre-M.E. brain for a few weeks to polish my writing style, I feel extremely lucky I've been able to do so much, bit by tiny bit. Some days I can still hardly read any of the site let alone believe I wrote so much of it!

It would not have been possible without all of you and the support, care, help, information, advice and time you've given me.

Thank you all so much,

Jodi Bassett, 2011

Acknowledgments

HFME would like to thank the following individuals for their significant contributions to HFME papers and/or helping to improve the HFME website and the HFME books:

- The late Aylwin Catchpole, a Canadian M.E. patient and patient advocate
- Lajla Mark, a Danish M.E. patient and patient advocate
- Peter, Lyn and Claire Bassett
- LK Woodruff, a US M.E. patient and patient advocate

HFME would like to thank the following individuals for their services as translators, translating some of the HFME's papers into other languages:

- Lotte Mayar
- Joan Steen
- Gitta Wolf
- Tatiana Coronado-Briceno

HFME would also like to thank all those who have contributed to HFME over the years. This includes everyone who has signed the guestbook, everyone who has sent us a case study, patient account or quote and everyone who has sent feedback, suggestions and comments by email.

Perhaps most of all HFME would like to thank the late Drs Melvin Ramsay and John Richardson and especially, Drs Elizabeth Dowsett and Byron Hyde. One could not put together an overview of the facts of M.E. without the valuable contributions of these individuals to the field of M.E. research. M.E. patients will be forever in their debt for all their research and truly heroic efforts toward M.E. advocacy. HFME thanks you all, most sincerely. Words cannot begin to express our gratitude for your work.

Thank you too to Dr Hyde for writing the wonderful foreword for this book.

Thank you also to all those doctors who have added some useful information to what we know about M.E. including Drs Paul Cheney, John Chia, Arnold Peckerman and A. Martin Lerner. Thanks also to all our fellow M.E. advocates who have helped to make some of the political facts of M.E. more accessible to us and to the general public.

Thank you to Professor Malcolm Hooper and Margaret Williams for such important papers as 'What is M.E.? What is CFS?' and 'The Mental Health Movement: Persecution of Patients?' Thanks from HFME also go to authors of excellent and uncompromising M.E. advocacy writings such as the late Gurli Bagnall, plus Cesar Quintero, LK Woodruff, Lajla Mark, The Committee for Justice and Recognition of M.E., Jill McLaughlin, Dr John Greensmith, and others.

HFME would also like to thank *everyone* who has helped, in big ways or small, to make others aware of HFME and the work of HFME. Every link made to our site or recommendation of one of our books helps get the facts out to more and more people, which is what it is all about.

The HFME reference list

All of the information concerning Myalgic Encephalomyelitis (M.E.) in this book and on the HFME website is fully referenced and has been compiled using the highest quality resources available, produced by the world's leading M.E. experts.

More experienced and more knowledgeable M.E. experts than these – Dr Byron Hyde and Dr Elizabeth Dowsett in particular – do not exist.

Between Drs Byron Hyde and Elizabeth Dowsett, and their mentors the late Drs John Richardson and Melvin Ramsay (respectively), these four doctors have been involved with M.E. research and M.E. patients for well over 100 years collectively, from the 1950s to the present day. Between them they have examined more than 15,000 individual (sporadic and epidemic) M.E. patients, as well as each authoring numerous studies and articles on M.E., and books (or chapters in books) on M.E. Again, more experienced, more knowledgeable and more credible M.E. experts than these simply do not exist.

The chapters in this book to a large extent provide merely a summary of the most important facts about M.E. They have been written for the benefit of those individuals without the time, inclination or ability to read each of the large number of far more complex and lengthy source papers. A full list of references is given at the end of each paper however, and these original references are highly recommended as essential additional reading for any physician (or anyone else) with a real interest in M.E.

Governments around the world are currently spending $0 a year on M.E. research. Considering the severity of the illness, and the vast numbers of patients involved, ranging in age from small children to adults, this is a worldwide disgrace. The fiction of 'CFS' represents outright medical fraud, involving serious medical abuse and neglect of patients, on a massive scale.

Not everyone was taken in by the 'CFS' insurance scam, however. A small but dedicated group of M.E. experts have each examined many thousands of M.E. patients and have made many remarkable discoveries about the pathology of M.E. These discoveries have confirmed many times over what was already known about M.E. prior to 1988, before M.E. research became tainted by irrelevant concepts of 'fatigue' and 'CFS' and then disappeared almost entirely.

Before reading the information in the reference list, please be aware of the following facts:

1. Myalgic Encephalomyelitis and 'Chronic Fatigue Syndrome' are not synonymous terms. The overwhelming majority of research on 'CFS' or 'CFIDS' or 'ME/CFS' or 'CFS/ME' or 'ICD-CFS' does not involve M.E. patients and is not relevant *in any way* to M.E. patients. However, if the M.E. community were to reject all 'CFS' labelled research as *only* relating to 'CFS' patients (including research which describes those abnormalities and characteristics unique to M.E. patients), this may support the myth that 'CFS' is just a 'watered down' definition of M.E. and that M.E. and 'CFS' are virtually the same thing and share many characteristics.

A very small number of 'CFS' studies refer in part to people with M.E. but it may not always be clear which parts refer to M.E. The <u>A warning on 'CFS' and 'ME/CFS' research and advocacy</u> paper is recommended reading and includes a checklist to help readers assess the relevance of individual 'CFS' studies to M.E. (if any) and explains some of the problems with this heterogeneous and skewed research.

In future, it is essential that M.E. research again be conducted using only M.E. defined patients and using only the term M.E. The bogus, financially-motivated disease category of 'CFS' must be abandoned.

2. The research referred to in the list below varies considerably in quality. Some is of a high scientific standard and relates wholly to M.E. and uses the correct terminology. Other studies or articles are included which may only have partial or minor possible relevance to M.E., use unscientific terms/concepts such as 'CFS,' 'ME/CFS,' 'CFS/ME,' 'CFIDS' or Myalgic 'Encephalopathy' and also include a significant amount of misinformation. Before reading this research it is also essential that the reader be aware of the most commonly used 'CFS' propaganda, as explained in <u>A warning on 'CFS' and 'ME/CFS' research and advocacy</u> and in more detail in <u>Putting research and articles on M.E. into context</u>.

The HFME reference list:

1. Acheson, AD 1954, *Encephalomyelitis associated with poliomyelitis virus,* Lancet: Nov 20th 1954:1044-1048
2. Acheson, AD 1956, *A New Clinical Entity?* THE LANCET. LONDON : May 26 1956:789-790
3. Acheson, AD 1955, *Outbreak at The Royal Free,* Lancet 20 August 1955:304-305
4. Acheson, AD 1959, *The Clinical Syndrome Variously Called Benign Myalgic Encephalomyelitis, Iceland Disease and Epidemic Neuromyasthenia,* Am J Med 1959:569 595 (Also: In *The Clinical and Scientific Basis of Myalgic Encephalomyelitis,* Hyde, Byron M.D. (ed) The Nightingale Foundation, Ottawa, pp. 129-158.)
5. Andreoletti L, Bourlet T, Moukassa D, et al. 2000, *Enteroviruses can persist with or without active viral replication in cardiac tissue of patients with end-stage ischemic or dilated cardiomyopathy,* J Infect Dis 2000;182:1222–7.
6. Archard LC, Bowles NE, Behan PO, Bell EJ, Doyle D 1988, <u>*Postviral fatigue syndrome: persistence of enterovirus RNA in muscle and elevated creatine kinase*</u>, Department of Biochemistry, Charing Cross and Westminster Medical School, London, J R Soc Med. 1988 Jun;81(6):326-9.
7. Baboonian C & Treasure T 1997 <u>*Meta-analysis of the association of enteroviruses with human heart disease,*</u> Heart,1997 Dec;78(6):539-43, Department of Cardiological Sciences, St George's Hospital Medical School, Tooting, London, UK.
8. Bagnall, Gurli 2006, *Civilization: Another word for barbarism* [Online], Available: http://www.hfme.org/wbagnall.htm
9. Bagnall, Gurli 2005, <u>*COERCION, CORRUPTION AND CONSPIRACY: A LETHAL MIX,*</u> [Online], Available: http://www.hfme.org/wbagnall.htm
10. Bagnall, Gurli 2005a, <u>*What is it About Psychiatry?*</u> [Online], Available: http://www.hfme.org/wbagnall.htm
11. Bagnall, Gurli 2004, *Myalgic Encephalomyelitis: A slender string to our bow* [Online], Available: http://www.hfme.org/wbagnallmasstob.htm
12. Bagnall, Gurli 2004a , <u>*COOKING THE BOOKS*</u> [Online], Available: http://www.hfme.org/wbagnall.htm
13. Bagnall, Gurli 2004b, <u>*PSYCHIATRY AND PERFIDY*</u> [Online], Available: http://www.hfme.org/wbagnall.htm
14. Bassett, Jodi 2010, *The Ultra-Comprehensive Myalgic Encephalomyelitis Symptom List* [Online], Available: http://www.hfme.org/themesymptomlist.htm
15. Bassett, Jodi 2010, *What is Myalgic Encephalomyelitis* [Online], Available: http://www.hfme.org/whatisme.htm
16. Bastien, Sheila PhD. 1992, *Patterns of Neuropsychological Abnormalities and Cognitive Impairment in Adults and Children* in Hyde, Byron M.D. (ed) 1992, *The Clinical and Scientific Basis of Myalgic Encephalomyelitis*, Nightingale Research Foundation, Ottawa
17. Behan, PO & Behan, WMH & Bell EJ 1985, *The postviral fatigue syndrome - an analysis of the findings in 50 cases*, Journal of Infection 1985:10:211-222.
18. Behan, BO & Behan, WMH 1988, *Postviral Fatigue Syndrome*, CRC Crit Rev Neurobiol 1988:4:2:157-178.
19. Behan PO, Behan WM, Gow JW, Cavanagh H & Gillespie S. 1993, <u>Enteroviruses and postviral fatigue syndrome,</u> Department of Neurology, University of Glasgow, UK, Ciba Found Symp. 1993;173:146-54; discussion 154-9.
20. Behan W, Gow JW & Curtis F 1999, *Blood Brain Barrier Breakdown in Myalgic Encephalomyelitis*, Presented at "Fatigue 2000" Conference, London 23rd 24th April 1999, arranged by The National ME Centre, Harold Wood, Essex, in conjunction with the Essex Neurosciences Unit
21. Bell, EJ & McCartney, RA. 1984, <u>A study of Coxsackie B virus infections 1972-1983</u>, J Hyg (Lond). 1984 Oct;93(2):197-203.
22. Bell EJ, McCartney RA & Riding MH 1988, <u>Coxsackie B viruses and myalgic encephalomyelitis</u>, Ruchill Hospital, Glasgow, J R Soc Med. 1988 Jun;81(6):329-31.
23. Bell, David S MD 1995, *The Doctor's Guide to CFIDS*, Perseus Books, Massachusetts
24. Bodian D, 1949, *Histopathological basis of findings*

in poliomyelitis, American Journal of Medicine, 1949;6:563-578

25. Bowles, NE et al. 1993, *Persistence of enterovirus RNA in muscle biopsy samples suggests that some cases of chronic fatigue syndrome result from a previous, inflammatory viral myopathy,* Journal of Medicine 1993:24: 2&3:145-180.

26. Bowles, M E, et al. 1986, *DETENTION OF COXSACKIE "B" VIRUS SPECIFIC RNA SEQUENCES IN MYOCARDIAL BIOPSY SAMPLES FROM PATIENTS WITH MYOCARDITIS AND DILATED CARDIOMYOPATHY,* Lancet, 1986; 1: 1120-1122

27. Brain, Lord 1962, *Diseases of the Nervous System,* Sixth Edition. Oxford University Press

28. Bruno, RL et al, 1996, *Polio Encephalitis and the Brain Generator Model of Post Viral Fatigue,* Journal of Chronic Fatigue Syndrome. 1996:2: (2,3):5 27

29. Bruno, RL, Frick NM, & Creange SJ et al. 1997, *A brain model for post viral fatigue syndrome,* ME Today, 1997;5/6:18-21

30. Bruno RL. et al. 1998, *Parallel between Post-Polio Fatigue and Chronic Fatigue Syndrome – A common Pathophysiology?,* American Journal of Medicine. 1998 105 (3A) : 66(s) – 73(s)

31. Bruno, RL 2002, *THE POLIO PARADOX,* Chapter 11: 164-166. Warner Books INC 2002, 1271 Avenue of the Americas, NY 10020.

32. Calder BD, Warnock PJ, McCartney RA, & Bell EJ 1987, <u>*Coxsackie B viruses and the post-viral syndrome: a prospective study in general practice,*</u> JRCGP 1987:37:11-14.

33. Carruthers, Bruce M. et al 2003, *Myalgic Encephalomyelitis/Chronic Fatigue Syndrome: Clinical Working Case Definition,* Haworth Medical Press, New York

34. Chabursky, Borys. Hyde, Byron M.D. & Anil Jain M.D. 1992, *A Description of Patients who Present with a Presumed Diagnosis of M.E.* in Hyde, Byron M.D. (ed) 1992, *The Clinical and Scientific Basis of Myalgic Encephalomyelitis,* Nightingale Research Foundation, Ottawa, pp. 19-24

35. Chapman NM, Kim KS, Drescher KM, Oka K, & Tracy S 2008, *5′ terminal deletions in the genome of a coxsackievirus B2 strain occurred naturally in human heart,* Virology 2008 Jun 5;375(2):480-91.

36. Cheney, Paul M.D. 2006, *The Heart of the Matter* [video recording], Available: http://www.hfme.org/wcheney.htm

37. Chia, John kai-sheng & Chia, Andrew Y, 2007, *Chronic fatigue syndrome is associated with chronic enterovirus infection of the stomach,* EV Med Research, United States, J Clin Pathol doi:10.1136/jcp.2007.050054

38. Chia JK 2005, <u>*The role of enterovirus in chronic fatigue syndrome,*</u> J Clin Pathol. 2005 Nov;58(11):1126-32, CEI Research Center, Torrance, CA 90505, USA.

39. Chia J, Chia A, Voeller M, Lee T, Chang R. 2010, <u>*Acute enterovirus infection followed by myalgic encephalomyelitis/chronic fatigue syndrome*</u> <u>*(ME/CFS) and viral persistence,*</u> J Clin Pathol. 2010 Feb;63(2):165-8. Epub 2009 Oct 14, EV Med Research, Torrance, California, USA.

40. Chia J, & Chia A. 2004, *Ribavirin and interferon a for the treatment of patients with chronic fatigue syndrome associated with chronic coxsackievirus B infection: a preliminary observation,* J Appl Research 2004;4:286–92.

41. Chia JK, Jou NS, & Majera L, et al. 2001, *The presence of enteroviral RNA (EV RNA) in peripheral blood mononuclear cells (PBMC) of patients with the chronic fatigue syndrome (CFS) associated with high levels of neutralizing antibodies to enteroviruses,* Abstract 405. Clin Infect Dis 2001;33:1157.

42. Chia J, & Chia A. 2002, *Detection of enteroviral RNA in the peripheral blood leukocytes of patients with the chronic fatigue syndrome,* Abstract 763. In: Program and abstracts of the 40th annual meeting of the Infectious Diseases Society of America. Chicago, IL: IDSA, 2002:178.

43. Clements GB, McGarry F, Nairn C & Galbraith DN 1995, Detection of enterovirus-specific RNA in serum: the relationship to chronic fatigue, J Med Virol. 1995 Feb;45(2):156-61, Regional Virus Laboratory, Ruchill Hospital, Glasgow, United Kingdom.

44. Colby J 2006, *Special problems of children with myalgic encephalomyelitis/chronic fatigue syndrome and the enteroviral link,* J Clin Pathol. 2007 Feb;60(2):125-8. Epub 2006 Aug 25, Tymes Trust, Ingatestone, Essex, UK.

45. Compston, Nigel D 1978, *An outbreak of encephalomyelitis in the Royal Free Hospital Group, London, in 1955,* Postgraduate Medical Journal 1978:54:722-724.

46. Cook DIB & Natelson BH et al. 2001, *Relationship of brain MRI abnormalities and physical functional status in chronic fatigue syndrome,* Int J Neurosci 2001:107: (1-2):1-6

47. Costa DC, Tannock C, & Brostoff J. 1995, Brainstem perfusion is impaired in patients with CFS, Quarterly Journal of Medicine. 1995;88:767-773

48. Crowley, N., Nelson, M., & Stovin, S. 1957, *Epidemiological aspects of an outbreak of encephalomyelitis at the Royal Free Hospital, London, in the summer of 1955.* Journal of Hygiene, 55, 102.

49. Cunningham L, Bowles NE & Archard LC 1991, *Persistent virus infection of muscle in postviral fatigue syndrome,* Br Med Bull. 1991 Oct;47(4):852-71, Department of Biochemistry, Charing Cross and Westminster Medical School, London, UK.

50. Cunningham, L & Lane, RJM & Archard, LC et al. 1990, *Persistence of enteroviral RNA in chronic fatigue syndrome is associated with the abnormal production of equal amounts of positive and negative strands of enteroviral RNA,* Journal of General Virology 1990:71:6:1399-1402.

51. Dalakas MC, 1995, Enterovirus and human muscular diseases, In: Human Enterovirus Infections, American Society for Microbiology, p. 387 – 398

52. Dalakas MC, 2003, Enteroviruses in chronic fatigue syndrome: "now you see them, now you don't", J Neurol Neurosurg Psychiatry, 2003;74:1361–1362

53. Dalldorf G, Sickles GM, Plager H, & Gifford R 1949, *A virus recovered from the faeces of "poliomyelitis" patients,* Pathogenic for suckling mice, Journal of Experimental Medicine. 1949; 89: 567-582

54. Daugherty, S. A., Henry, B. E., Peterson, D. L., Swarts, R. L., Bastien, S., Thomas, R. S., et al. 1991, *Chronic fatigue syndrome in northern Nevada,* Reviews of Infectious Diseases, 13(Suppl. 1), S39-S44.

55. De La Torre JC, Mallory M, & Brot M 1996, Viral persistence in neurons alters synaptic plasticity and cognition functions without destruction of brain cells. Virology. 1996;220:508-515

56. Devanur LD, & Kerr JR 2006, *Chronic fatigue syndrome,* J Clin Virol 2006;37:139–50.

57. Douche-Aourik F et al. 2003, *Detection of enterovirus in human skeletal muscle from patients with chronic inflammatory muscle disease in fibromyalgia (CFS) and healthy subjects,* Journal of Medical virology 2003:71:540-547.

58. Dowsett, Elizabeth MBChB. 2002a, *The impact of persistent enteroviral infection,* [Online], Available: http://www.hfme.org/wdowsett.htm

59. Dowsett, Elizabeth MBChB. 2002b, *MEDICAL RESEARCH COUNCIL DRAFT DOCUMENT FOR PUBLIC CONSULTATION,* [Online], Available: http://www.hfme.org/wdowsett.htm

60. Dowsett, Elizabeth MBChB. 2001a, *THE LATE EFFECTS OF ME Can they be distinguished from the Post-polio syndrome?* [Online], Available: http://www.hfme.org/wdowsett.htm

61. Dowsett, Elizabeth MBChB. 2001b, *A rose by any other name* [Online], Available: http://www.hfme.org/wdowsett.htm

62. Dowsett, Elizabeth MBChB. 2000, *Mobility problems in ME* [Online], Available: http://www.hfme.org/wdowsett.htm

63. Dowsett, Elizabeth MBChB. 1999a, *Redefinitions of ME* [Online], Available: http://www.hfme.org/wdowsett.htm

64. Dowsett, Elizabeth MBChB. 1999b, *Research into ME 1988 - 1998 Too much PHILOSOPHY and too little BASIC SCIENCE!,* [Online], Available: http://www.hfme.org/wdowsett.htm

65. Dowsett EG & Richardson J 1999c, The Epidemiology of Myalgic Encephalomyelitis (ME) in the UK, Evidence submitted to the All Party Parliamentary Group of Members of Parliament, 23 Nov 1999

66. Dowsett, Elizabeth MBChB. 1998, *Can Hysteria be diagnosed with confidence ? - Conflicts in British Research,* [Online], Available: http://www.hfme.org/wdowsett.htm

67. Dowsett, EG 1998a, *Enteroviral Infections and their Sequelae,* BSAEM; RCGP 1998:1-10.

68. Dowsett EG & DM Jones DM 1998a, The Organic Basis of ME / CFS, Information and Statistics presented to the Chief Medical Officer in person at a meeting on 11th March 1998

69. Dowsett E, 1988, *Human enteroviral infections,* Journal of Hospital Infection 1988:11:103-115.

70. Dowsett, EG & Colby, J 1997, *Long Term Sickness Absence due to ME/CFS in UK Schools - an epidemiological study with medical and educational implications,* Journal of Chronic Fatigue Syndrome 1997; 3 (2): 2942

71. Dowsett, Elizabeth MBChB. n.d. a, *Differences between ME and CFS,* [Online], Available: http://www.hfme.org/wdowsett.htm

72. Dowsett, Elizabeth MBChB. n.d. b, *Time to put the exercise cure to rest,* [Online], Available: http://www.hfme.org/wdowsett.htm

73. Dowsett, Elizabeth MBChB. n.d. c, *Is Stress more than a modern buzz word?,* [Online], Available: http://www.hfme.org/wdowsett.htm

74. Dowsett, Elizabeth MBChB. n.d. d, *Brain problems in ME - is there a simple explanation?* , [Online], Available: http://www.hfme.org/wdowsett.htm

75. Dowsett, Elizabeth MBChB. in: Colby, Jane 1996, *ME: The New Plague,* Ipswitch Book Company, Ipswitch.

76. Dowsett E. & Ramsay, A.M. n.d., ' *Myalgic Encephalomyelitis: Then and Now' The Clinical and Scientific Basis of Myalgic Encephalomyelitis,* B. Hyde (ed.), The Nightingale Foundation, Ottawa, pp. 81-84.

77. Dowsett E., Ramsay A.M., McCartney A.R., & Bell E.J. 1990, ' *Myalgic Encephalomyelitis: A persistent Enteroviral Infection?' in The Clinical and Scientific Basis of Myalgic Encephalomyelitis,* B. Hyde (ed.), The Nightingale Foundation, Ottawa, pp. 285-291.

78. Dunn, Linda (in consultation with the Cross Party Group on ME) 2005,*Myalgic Encephalomyelitis. The impact on sufferers: is health policy in Scotland on the right path?* [Online], Available: (Link)

79. Fegan KG, Behan PO, & Bell EJ 1983, *Myalgic encephalomyelitis--report of an epidemic,* J R Coll Gen Pract. 1983 Jun;33(251):335-7.

80. Frisk G. 2005, *Mechanism of chronic enteroviral persistence in tissue,* Curr Opin Infect Dis 2001;14:251–6.

81. Galbraith DN et al, 1997, *Evidence for Enteroviral Persistence in Humans,* J Gen Virol 1997:78:307-312

82. Galbraith, DN et al.1995, *Phylogenetic analysis of short enteroviral sequences from patients with chronic fatigue syndrome,* Journal of General Virology 1995:76:1701-1707.

83. Galbraith DN, Nairn C, & Clements GB. 1995, *Phylogenetic analysis of short enteroviral sequences from patients with chronic fatigue syndrome,* J Gen Virol 1995;76:1701–7.

84. Gilliam, AG 1934, *Epidemiological study of an epidemic diagnosed as poliomyelitis occurring*

among the personnel of Los Angeles County General Hospital during the summer of 1934, Public Health Bulletin, US Treasury Department No. 240, 1938

85. Gow, JW et al. 1991, *Enteroviral RNA sequences detected by polymerase chain reaction in muscles of patients with postviral fatigue syndrome,* BMJ 1991:302:696-696.

86. Gow, JW & Behan, WM 1991, <u>*Amplification and identification of enteroviral sequences in the postviral fatigue syndrome*</u>, Br Med Bull. 1991 Oct;47(4):872-85, University Department of Neurology, Southern General Hospital, Glasgow, UK.

87. Gow, JW & Behan, WM et al. 1994, *Studies on enterovirus in patients with CFS,* Clin Infect Dis 18, S126-S129

88. Gray, JA 1984, <u>*Some long-term sequelae of Coxsackie B virus infection*</u>, J R Coll Gen Pract, Jan;34(258):3-5.

89. Griffin, Susan 2000, From *What Her Body Thought: A Journey into the Shadows* in *Stricken; Voices from the Hidden Epidemic of Chronic Fatigue Syndrome,* ed Peggy Munson, The Haworth Press, New York

90. Grufferman, S & Komaroff, AL & Bell, DS & Peterson, DL & Daugherty, S & Bastien, S et al. 1991, Considerations in the Design of Studies of Chronic Fatigue Syndrome, Reviews of Infectious Diseases, Volume 13, Supplement 1: S1 - S140. University of Chicago Press.

91. Grufferman, S. 1992, *Epidemiologic and immunologic findings in clusters of chronic fatigue syndrome,* In *The Clinical and Scientific Basis of Myalgic Encephalomyelitis,* Hyde, Byron M.D. (ed) The Nightingale Foundation, Ottawa, pp. 189-195

92. Henderson, D. A., & Shelokov, A. 1992, *Epidemic neuromyasthenia-clinical syndrome,* In *The Clinical and Scientific Basis of Myalgic Encephalomyelitis,* Hyde, Byron M.D. (ed) The Nightingale Foundation, Ottawa, pp. 159-175.

93. Hooper, M. & Marshall E.P. & Williams M. 2007, *Corporate Collusion?,* Available: http://www.hfme.org/whooper.htm

94. Hooper, M. & Marshall E.P. 2005a, *Illustrations of Clinical Observations and International Research Findings from 1955 to 2005 that demonstrate the organic aetiology of Myalgic Encephalomyelitis* [Online], Available: http://www.hfme.org/wmarwillhoopgibsonenqui.htm

95. Hooper, M. & Marshall E.P. 2005b, *Myalgic Encephalomyelitis: Why no accountability?* [Online], Available: http://www.hfme.org/wmarwillmewna.htm

96. Hooper, M. & Montague S 2001a, *Concerns about the forthcoming UK Chief medical officer's report on Myalgic Encephalomyelitis (ME) and Chronic Fatigue Syndrome (CFS) notably the intention to advise clinicians that only limited investigations are necessary* (The Montague/Hooper paper) [Online], Available: http://www.hfme.org/whooper.htm

97. Hooper, M. & Montague S 2001b, *Concepts of accountability* [Online], Available: http://www.hfme.org/whooper.htm

98. Hooper, M 2006, Myalgic Encephalomyelitis (ME): a review with emphasis on key findings in biomedical research J. Clin. Pathol. published online 25 Aug 2006; doi:10.1136/jcp.2006.042408 Available: http://www.hfme.org/whooper.htm

99. Hooper, M. 2003a, *The MENTAL HEALTH MOVEMENT: PERSECUTION OF PATIENTS?* [Online], Available: http://www.hfme.org/whooper.htm

100. Hooper, M 2003b, *Engaging with M.E.: Towards Understanding, Diagnosis and Treatment,* University of Sunderland, UK

101. Hooper, M. n.d., *The Terminology of ME and CFS* [Online], Available: http://www.hfme.org/whooper.htm

102. Hooper, M. Marshall E.P. & Williams, M. 2001, *What is ME? What is CFS? Information for Clinicians and Lawyers,* [Online], Available: http://www.hfme.org/wmarwillhoopwimewicfs.htm

103. Hyde, Byron M.D. 2009, *Missed Diagnoses: M.E. and CFS,* Nightingale Research Foundation, Canada

104. Hyde, Byron M.D. 2007, *The Nightingale Definition of Myalgic Encephalomyelitis* [Online], Available: www.hfme.org/whyde.htm

105. Hyde, Byron M.D. 2006, *A New and Simple Definition of Myalgic Encephalomyelitis and a New Simple Definition of Chronic Fatigue Syndrome & A Brief History of Myalgic Encephalomyelitis & An Irreverent History of Chronic Fatigue Syndrome* [Online], Available: http://www.hfme.org/whyde.htm

106. Hyde, Byron M.D. 2003, *The Complexities of Diagnosis* in (ed) Jason, Leonard at et al. 2003 *Handbook of Chronic Fatigue Syndrome* by Ross Wiley and Sons, USA. Available: http://www.hfme.org/whyde.htm

107. Hyde, Byron M.D. 1992, *Preface* in Hyde, Byron M.D. (ed) 1992, *The Clinical and Scientific Basis of Myalgic Encephalomyelitis,* Nightingale Research Foundation, Ottawa.

108. Hyde, B. M. 1992b, A bibliography of ME epidemics, In *The Clinical and Scientific Basis of Myalgic Encephalomyelitis,* Hyde, Byron M.D. (ed) The Nightingale Foundation, Ottawa, pp. 176-186.

109. Hyde, B. M., Biddle, R., & McNamara, T. 1992, Magnetic resonance in the diagnosis of ME/CFS, a review, In *The Clinical and Scientific Basis of Myalgic Encephalomyelitis,* Hyde, Byron M.D. (ed) The Nightingale Foundation, Ottawa, pp. 425-431.

110. Hyde, Byron M.D. & Anil Jain M.D. 1992, *Clinical Observations of Central Nervous System Dysfunction in Post Infectious, Acute Onset M.E.* in Hyde, Byron M.D. (ed) 1992, *The Clinical and Scientific Basis of Myalgic Encephalomyelitis,* Nightingale Research Foundation, Ottawa, pp. 38-65.

111. Hyde, Byron M.D. & Anil Jain M.D. 1992a, *Cardiac and Cardiovascular aspects of M.E.: A Review* in Hyde, Byron M.D. (ed) 1992, *The*

Clinical and Scientific Basis of Myalgic Encephalomyelitis, Nightingale Research Foundation, Ottawa, pp. 375-383.

112. Hyde, Byron M.D., Bastien S Ph.D. & Anil Jain M.D. 1992, *General Information, Post Infectious, Acute Onset M.E.* in Hyde, Byron M.D. (ed) 1992, *The Clinical and Scientific Basis of Myalgic Encephalomyelitis*, Nightingale Research Foundation, Ottawa, pp. 25-37.

113. Hyde, Byron M.D. 1988, *Are Myalgic Encephalomyelitis and CFS Synonymous Terms?* [Online], Available: http://www.hfme.org/whyde.htm

114. Hyde, Byron MD. 2010, *Mental health problems in patients with myalgic encephalomyelitis and fibromyalgia syndrome,* [Online], Available: http://www.hfme.org/whyde.htm

115. Johnson, Hillary 1996, *Osler's Web*, Crown Publishers, New York

116. Jones, Doris M. M.S.c. 1998, *SOME FACTS AND FIGURES ON CBT, GET AND OTHER APPROACHES Directly from the 'Horses' Mouths* [Online]. Available: link

117. Innes, AGB 1970, Encephalomyelitis resembling benign myalgic encephalomyelitis, Lancet 1970: 969-971

118. Jou N-T et al. 2001, *Enterovirus in Chronic Fatigue Syndrome* Arthritis & Rheumatism 2001:44:S9:S351.

119. Keighley BD & Bell EJ, 1983, *Sporadic myalgic encephalomyelitis in a rural practice,* JRCP 1983:33:339-341.

120. Kerr, JR 2007, *Enterovirus infection of the stomach in Chronic Fatigue Syndrome / Myalgic Encephalomyelitis,* J Clin Pathol. 2007 Sep 14, St George's University of London, United Kingdom.

121. Kuhl U, Pauschinger M, & Seeberg B, et al. 2005, *Viral persistence in the myocardium is associated with progressive cardiac dysfunction*, Circulation 2005;112:1965–70.

122. Lane, RJM & Archard, LC et al. 2003, *Enterovirus related metabolic myopathy: a postviral fatigue syndrome*, JNNP 2003:74:1382-1386.

123. Lerner AM et al, 1997, *Cardiac involvement in patients with CFS as documented with Holter monitor and biopsy data,* Infectious Diseases in Clinical Practice 1997:6:327-333

124. Lerner, MA & Corning, PD 1998, *RESEARCH BREAKTHROUGH: ME/CFS AN INFECTIOUS CARDIOMYOPATHY?* [Online], Available: http://www.hfme.org/wlerner.htm

125. Levine, Susan 2001, *Prevalence in the Cerebrospinal Fluid of the Following Infectious Agents in a Cohort of 12 CFS Subjects: Human Herpes Virus-6 and 8; Chlamydia Species; Mycoplasma Species, EBV; CMV and Coxsackievirus ,* JCFS 2002:9:1/2:41-51.

126. Livingstone, Churchill 1991, *Postviral Fatigue Syndrome*, British Medical Bulletin, 1991:47:4: 793-907.

127. Lyle, WH & Chamberlain, RN, 1978, *Epidemic Neuromyasthenia 1934 1977. current approaches,*

Postgraduate Medical Journal 1978:54:637:705 774 pub: Blackwell Scientific Publications, Oxford

128. Lyle, WH 1959, *An outbreak of a disease believed to have been caused by ECHO 9 Virus*, Annals of Internal Medicine 1959; 51: 248-269

129. Mar, Countess of. 2004, *House of Lords Debate* [Online], Available: (link in title)

130. Marshall, Eileen & Williams, Margaret. 2006a, *Some of the abnormalities that have been demonstrated in ME/CFS* [Online], Available: http://www.hfme.org/wmarwillsotathbdime.htm

131. Marshall, Eileen & Williams, Margaret. 2006b, *Inquest implications* [Online], Available: http://www.hfme.org/wmarwillinquest.htm

132. Marshall, Eileen & Williams, Margaret. 2006c, *Myalgic Encephalomyelitis exists: True or false?* [Online], Available: http://www.hfme.org/wmarwillinquest.htm

133. Marshall, Eileen & Williams, Margaret. 2005a, *Problems and solutions?* [Online], Available: http://www.hfme.org/wmarwillpas.htm

134. Marshall, Eileen & Williams, Margaret. 2005b, *Proof positive? Evidence of the deliberate creation via social constructionism of "psychosocial" illness by cult indoctrination of State agencies, and the impact of this on social and welfare policy* [Online], Available: http://www.hfme.org/wmarwillpp.htm

135. Marshall, Eileen & Williams, Margaret. 2005c, *To set the record straight about Ean Proctor from the Isle of Man* [Online], Available: http://www.hfme.org/wmarwilltstrsaep.htm

136. Marshall, Eileen & Williams, Margaret. 2005d, *Profits before Patients?* [Online], Available: http://www.hfme.org/wmarwillpbp.htm

137. Marshall, Eileen & Williams, Margaret. 2004a, *A note on the term Myalgic Encephalomyelitis* [Online], Available: http://www.hfme.org/wmarwillnottme.htm

138. Marshall, Eileen & Williams, Margaret. 2004b, *Transparency in government* [Online], Available: http://www.hfme.org/wmarwilltig.htm

139. Martin, JW 1989, *Detection of Viral Related Sequences in CFS Patients using the Polymerase Chain Reaction*, The Nightingale Research Foundation, 1989: 1-5

140. McCartney RA, Banatvala JE & Bell EJ 1986, *Routine use of mu-antibody-capture ELISA for the serological diagnosis of Coxsackie B virus infections ,* J Med Virol. 1986 Jul;19(3):205-12.

141. McGarry F, Gow J and Behan PO, 1994, *Enterovirus in the chronic fatigue syndrome,* Ann Intern Med 1994:120:972 973

142. Melnick JL. Ledinko N, Kaplan A, & Kraft E. 1950, *Ohio Strains of a Virus Pathogenic for Infant Mice (Coxsackie Group). Simultaneous occurrence with poliomyelitis virus in patients with "summer grippe"*, Journal of Experimental Medicine. 1950 : 91:185-195

143. Mena, I (1991, *Study of cerebral perfusion by neuro-SPECT in patients with chronic fatigue syndrome*, Presented at Chronic Fatigue Syndrome:

Current Theory and Treatment conference, Bel Air, CA.

144. Michell, Lynn 2003, *Shattered: Life with ME,* Thorsons Publishers, London

145. Mocé-Llivina L, Lucena F & Jofre J, *Enteroviruses and bacteriophages in bathing waters,* Appl Environ Microbiol. 2005 Nov;71(11):6838-44, Department of Microbiology, Faculty of Biology, University of Barcelona, Avda. Diagonal, 645 Edifici Annex, Planta 0, E-08028 Barcelona, Spain.

146. Montague, T.J., Marrie, T., Klassen, G. Bewick, D., & Horacek, B.M. 1989, *Cardiac Function at Rest and with Exercise in the Chronic Fatigue Syndrome,* April 1989, Chest, Vol 95, p779-784,.

147. Muir P et al. 1998, *Molecular typing of enteroviruses: current status and future requirements,* The European Union Concerted Action on Virus Meningitis and Encephalitis, Clin Microbiol Rev. 1998 Jan;11(1):202-27, Department of Virology, United Medical School of Guy's Hospital, London, United Kingdom.

148. Muir, P & Archard, LC 1994, *There is evidence for persistent enterovirus infection in chronic medical conditions in humans,* Reviews in Medical Virology, 1994; 4: 245-250

149. Munson, Mary 2000, *Taking the Rap: Parents, Blame and Pediatric CFIDS* in Stricken; Voices from the Hidden Epidemic of chronic Fatigue Syndrome, ed Peggy Munson, The Haworth Press, New York

150. Munson, Peggy (ed) 2000, *Stricken; Voices from the Hidden Epidemic of Chronic Fatigue Syndrome,* The Haworth Press, New York

151. Murdoch JC 1988, *The Myalgic Encephalomyelitis Syndrome,* Family Practice 1988:5:4:302 306. pub: Oxford University Press

152. Myhill, S & Booth, Norman E & McLaren-Howard, John 2009, *Chronic fatigue syndrome and mitochondrial dysfunction,* International Journal of Clinical and Experimental Medicine (IJCEM) (ISSN 1940-5901) 2, 1-16

153. Nairn, C et al. *Comparison of Coxsackie B Neutralisation and Enteroviral PCR in Chronic Fatigue Patients,* Journal of Medical Virology 1995:46:310-313.

154. Natelson BH et al. 2005, *Spinal fluid abnormalities in patients with chronic fatigue syndrome,* CFS Cooperative Research Center and Department of Neurosciences, University of Medicine and Dentistry of New Jersey-New Jersey Medical School, Newark, New Jersey, USA.

155. Oberste, M.S. & Pallansch, M 2003, *Establishing evidence for enterovirus infection in chronic diseases,* Ann NY Acad Sci 2003;1005:23–31.

156. Patterson, J et al. 1995, *SPECT Brain Imaging in Chronic Fatigue Syndrome,* EOS - J Immunol Immunpharmacol 1995:vol 15: no.1-2:53-58

157. Peckerman A, LaManca JJ, Dahl KA, Chemitiganti R, Qureishi B, Natelson BH. 2003, *Abnormal Impedance Cardiography Predicts Symptom Severity in Chronic Fatigue Syndrome,* The American Journal of the Medical Sciences: 2003:326:(2):55-60)

158. Pellew RAA, 1955, *A clinical description of a disease resembling poliomyelitis seen in Adelaide,* Med J Aust 1955:42.480 482

159. Poser, Charles , 1992, *'The Differential Diagnosis between Multiple Sclerosis and Chronic Fatigue Postviral Syndrome'* The Clinical and Scientific Basis of Myalgic Encephalomyelitis, Hyde, Byron M.D. (ed) The Nightingale Foundation, Ottawa

160. Preedy VR, Smith DG, Salisbury JR & Peters TJ 1993, *Biochemical and muscle studies in patients with acute onset post-viral fatigue syndrome,* J Clin Pathol. 1993 Aug;46(8):722-6, Department of Clinical Biochemistry, King's College School of Medicine & Dentistry, London

161. Quintero, Sezar 2002, *Sophisticated Investigation, or How to Disguise a Disease,* [Online], Available: http://www.geocities.com/sezar99q/

162. Ramsay, A. 1988, *Myalgic Encephalomyelitis and Postviral Fatigue States: The saga of Royal Free Disease,* Gower Medical Publishing, London.

163. Ramsay, AM 1988, *Myalgic encephalomyelitis or what?* Lancet 1988:100

164. Ramsay, AM & Dowsett, EG et al, 1977, *Icelandic Disease (Benign Myalgic Encephalomyelitis or Royal Free Disease),* BMJ May 1977:1350

165. Ramsay, Melvin A. 1989, ME Association Newsletter, Winter 1989: pp. 20-21.

166. Ramsay, Melvin A. n.d., *The Myalgic Encephalomyelitis syndrome* [Online], Available: http://www.hfme.org/wramsay.htm

167. Ramsay, Melvin A. 1986, *MYALGIC ENCEPHALOMYELITIS : A Baffling Syndrome With a Tragic Aftermath.* [Online], Available: http://www.hfme.org/wramsay.htm

168. Ramsay, AM. 1956, Encephalomyelitis simulating polio myelitis. Lancet. 1956;1: 761-766

169. REETOO K N. et al. 2000, *QUANTATIVE ANALYSIS OF VIRAL RNA KINETICS IN COXSACKIE VIRUS B3-INDUCED MAURINE MYOCARDITIS,* Journal of General Virology. 2000; 81: 2755-2762.

170. Richardson, J. 1999, *Myalgic Encephalomyelitis: Guidelines for doctors* [Online], Available: http://www.hfme.org/wrichardson.htm

171. Richardson, J & Dowsett EG, 1999, *The Epidemiology of Myalgic Encephalomyelitis (ME) in the UK.* Evidence submitted to the All Party Parliamentary Group of Members of Parliament, 23 Nov 1999

172. Richardson, John 2001, *Viral Isolation from Brain in Myalgic Encephalomyelitis (A Case Report),* Journal of CFS 2001:9: (3-4):15-19

173. Richardson, J. n.d., *'M.E., The Epidemiological and Clinical Observations of a Rural Practitioner,'* The Clinical and Scientific Basis of Myalgic Encephalomyelitis, Hyde, Byron M.D. (ed) The Nightingale Foundation, Ottawa, pp. 85-94.

174. Richardson, John 2001, *Enteroviral and Toxin Mediated Myalgic Encephalomyelitis and Other*

Organ Pathologies, The Haworth Press Inc. New York

175. Richardson, J. 1995, *DISTURBANCE OF HYPOTHALMIC FUNCTION AND EVIDENCE FOR PERSISTENT ENTEROVIRUS INFECTION IN PATIENTS WITH CHRONIC FATIGUE SYNDROME,* JCFS 1995; 1 (2): 623-624.

176. Richardson. J, Costa D C. 1998, *RELATIONSHIP BETWEEN SPECT SCANS AND BUSPIRONE TESTS IN PATIENTS WITH ME/CFS,* JCFS 1998; 4 (3): 23-38

177. Rotbart H A. 2000, *ANTIVIRAL EFFECT OF COMBINATION OF ENVIROXIME AND DISOXARTIL ON COXSACKIE VIRUS B1 INFECTION,* Acta Virologica. 2000;44 (2): 73-78.

178. Rotbart HA, O'Connel JF, & McKinlay MA 1998, *Treatment of Human Enterovirus Infection,* Antiviral Research. 1998; 38:1-14.

179. Rotholz, James 2000, *CFIDS, Suffering and the Divine* in *Stricken; Voices from the Hidden Epidemic of Chronic Fatigue Syndrome,* ed Peggy Munson, The Haworth Press, New York

180. Rutherford, Jonathan 2007, *New Labour, the market state, and the end of welfare* [Online] Available: http://www.lwbooks.co.uk/journals/articles/rutherfo rd07.html

181. Ryll, Erich D. 1994, INFECTIOUS VENULITIS, CHRONIC FATIGUE SYNDROME, MYALGIC ENCEPHALOMYELITIS, [Online], Available: http://www.geocities.com/tcjrme/recommend23.htm l

182. San Diego Chronic Diseases, [Online], Available: http://www.sandiegocd.org/index.html

183. Satish R Raj, MD M.S.CI 2006, *The Postural Tachycardia Syndrome (POTS): Pathophysiology, Diagnosis & Management,* Indian Pacing Electrophysiol J. 2006 Apr-Jun; 6(2): 84-99.

184. Satoh M, Tamura G & Segawa I, et al. 1996, *Expression of cytokine genes and presence of enteroviral genomic RNA in endomyocardial biopsy tissues of myocarditis and dilated cardiomyopathy ,* Virchows Arch. 1996 Feb;427(5):503-9.

185. Schwartz RB & Komaroff AL et al. 1994, *Detection of Intracranial Abnormalities in Patients with Chronic Fatigue Syndrome: comparison of MR imaging and SPECT,* Am J Roentgenol 1994:162:935-941.

186. Selwyn S & Howiti LF 1962, *A Mosaic of Enteroviruses. Polio, Coxsackie and Echo infection in a group of families,* The Lancet, 1962; 2: 548-551

187. Shelokov, A., Habel, K., Verder, E., & Welch, W. 1957, *Epidemic neuromyasthenia: An outbreak of poliomyelitis-like illness in student nurses,* New England Journal of Medicine, 257, 345.

188. Sigurdsson B, Sigurjonsson J & Sigurdsson J 1950, *Disease epidemic in Iceland simulating poliomyelitis,* American Journal of Hygiene, 52, 222.

189. Southern P, & Oldstone MBA. 1986, *Medical consequences of persistent viral infection,* New England Journal of Medicine : 1986; 314 : 359-367

190. Streeten, David H. P. 1987, *Orthostatic Disorders of the Circulation, Mechanisms, Manifestations, and Treatment,* Plenum Medical Book Company, New York and London.

191. The 25% M.E. Group 2005, *Submission To The Parliamentary Inquiry Into Progress In The Scientific Research Of M.E. By The 25% ME Group,* [Online]. Available: http://www.hfmc.org/w25group.htm

192. The 25% M.E. Group. 2004, *SEVERELY AFFECTED ME (MYALGIC ENCEPHALOMYELITIS) ANALYSIS REPORT ON QUESTIONNAIRE* [Online]. Available: http://www.hfme.org/w25group.htm

193. The 25% M.E. Group n.d.a., *25% ME Group Response to the CFS/ME Research Advisory Group's Draft Report,* [Online]. Available: http://www.hfme.org/w25group.htm

194. The 25% M.E. Group n.d.b., *PRESS RELEASE & GENERAL STATEMENT,* [Online]. Available: http://www.hfme.org/w25group.htm

195. The 25% M.E. Group and MERGE 2002, *Severely Overlooked by Science - An Overview of Research on Severely-ill People with ME,* [Online]. Available: http://www.hfme.org/w25group.htm

196. The 25% M.E. Group and MERGE 2000, *Survey of the experiences of housebound/bed-bound ME patients* [Online] Available: http://www.meresearch.org.uk/research/reviews/exp eriences.html

197. The Committee for Justice and Recognition of Myalgic Encephalomyelitis 2007, [Online], Available: http://www.geocities.com/tcjrme/

198. TCJRME 2010, Recent Epidemics: Why are the Epidemics so important, [Online], Available: http://www.hfme.org/wtjcjrme.htm

199. TCJRME 2010, A Public Statement to Government Health Ministers and an ALERT to citizens worldwide [Online], Available: http://www.hfme.org/wtjcjrme.htm

200. TCJRME 2010, ME and CFS: The Definitions, [Online], Available: http://www.hfme.org/wtjcjrme.htm

201. The ME Society of America website 2007, [Online], Available: http://www.cfids-cab.org/MESA/framework.html

202. TIRELLI, V. et al. 1998, *Brain positron emission tomography (PET) in Chronic Fatigue Syndrome,* American Journal of Medicine, 1998; 105 (3a) : 54(s) – 58(s).

203. Tracy, Steven & Chapmen, 2009, Nora M, *Human Enteroviruses and Chronic Infectious Disease,* Journal of IiME, [Online], Available: link

204. Ueno, H &Yokota, Y, et al. 1995, *Significance of detection of enterovirus RNA in myocardial tissues by reverse transcription-polymerase chain reaction,* Int J Cardiol. 1995 Sep;51(2):157-64, Department of Internal Medicine, Hyogo Medical Center for Adults, Akashi, Japan.

205. Verillo, Erica F & Gellman, Lauren M 1997, *CFIDS - A Treatment Guide,* St. Martin's Griffin, New York

206. Walker, Martin J. 2003, *SKEWED: Psychiatric Hegemony and the Manufacture of Mental Illness in Multiple Chemical Sensitivity, Gulf War Syndrome, Myalgic Encephalomyelitis and Chronic Fatigue Syndrome* , Slingshot Publications, London.

207. Wallis, AL 1957, *An investigation into an unusual disease in epidemic and sporadic form in general practice in Cumberland in 1955 and subsequent years,* University of Edinburgh Doctoral Thesis 1957

208. Warner CL, Heffner, RR & Cookfair, D 1992, *Neuromuscular Abnormalities in Patients with Chronic Fatigue Syndrome,* In: The Clinical and Scientific Basis of Myalgic Encephalomyelitis, Hyde, Byron M.D. (ed) The Nightingale Foundation, Ottawa

209. Watson WS, McCreath GT, Chaudhuri A and Behan PO 1997, *A Possible Cell Membrane Transport Defect in Chronic Fatigue Syndrome?,* JCFS 1997 page 1-13

210. Web M.D. website 2009, [Online], Available: http://www.webmd.com

211. Williams, Margaret 2007, *Wessely, Woodstock and Warfare,* [Online], Available: http://www.hfme.org.htm

212. Williams, Margaret 2004, *Critical considerations,* [Online], Available: http://www.hfme.org/wmarwillcc.htm

213. Williams, Margaret 2003, *Quotations from "SOMATIC MEDICINE ABUSES PSYCHIATRY - AND NEGLECTS CAUSAL RESEARCH" by Per Dalen* [Online], Available: http://www.hfme.org/wmarwillmwrqf.htm

214. Wilson, Lawrence M.D. 2010, *Hypoglycemia,* [Online], Available: http://drlwilson.com/ARTICLES/HYPOGLYCEMIA.htm

215. Yousef, G, Mowbray, J et al. 1988, *Chronic Enterovirus Infection in Patients with Postviral Fatigue Syndrome,* The Lancet, Jan 23, 1988 i: 146-50.

216. Zhang H, Li Y, McClean DR & Richardson PJ, et al. 2004, *Detection of enterovirus capsid protein VP1 in myocardium from cases of myocarditis or dilated cardiomyopathy by immunohistochemistry: further evidence of enterovirus persistence in myocytes,* Med Microbiol Immunol. 2004 May;193(2-3):109-14. Epub 2003 Nov 22, Cell and Molecular Biology Section, Division of Biomedical Sciences, Faculty of Medicine, Imperial College, London, UK.

Note that while HFME supports the extremely important work being done by the Nightingale Foundation and Dr Byron Hyde as well as work done by Dr Elizabeth Dowsett and others, views expressed in this book are not their responsibility and are the sole responsibility of the listed authors.

Further quotes

'Although the authors of these definitions have repeatedly stated that they are defining a syndrome and not a specific disease, patient, physician, and insurer alike have tended to treat this syndrome as a specific disease or illness, with at times a potentially specific treatment and a specific outcome. This has resulted in much confusion. The physician and patient alike should remember that CFS is not a disease. It is a chronic fatigue state. Patients who conform to any of these CFS definitions may still have an undiagnosed major illness, certain of which are potentially treatable.'
DR BYRON HYDE 2003

'With the rapid development of technology, the UK retained its reputation as a leading centre of M.E. research and remained able to report clinical studies backed up by molecular biology, brain imaging, sophisticated hormonal and other biochemical studies. At this point, with sound evidence of an infective cause, the way in which such infection is spread and the pathogenisis of the disease, why were we urged to adopt the "fatigue definitions" inflicted upon M.E. sufferers by USA scientists?'
REDEFINITIONS OF M.E. - A 20TH CENTURY PHENOMENON BY DR ELIZABETH DOWSETT

'The incubation period from time of contact with the infection until the appearance of the illness is approximately 4-7 days. In its epidemic form M.E. was most commonly seen in (a) Health Care Workers, (b) children and older students in residential schools, nurses residences and hospitals, (c) in military barracks where students or soldiers were housed in close proximity further supporting the belief in its infectious nature. Although M.E. was not caused by poliovirus in the Akureyri epidemic, infection with M.E. somehow protected the patients from the polio epidemic that swept though Iceland in the 1950s. Polioviruses represent three of approximately 100 different enteroviruses. This was the reason why many in the UK believed that some of these epidemics were probably caused by a less lethal non-polio form of enteroviruses such as ECHO, Coxsackie, the numbered and new enteroviruses.'
DR BYRON HYDE

''To suggest that M.E. is merely one subgroup amongst this heterogenous collection of physiological and pathological states, thus making any attempt at differential diagnosis between them impossibly expensive to pursue; to suggest that diagnosis must be delayed for 6 months, vitiates any real attempt at virus investigation, especially among the young.'
REDEFINITIONS OF M.E. - A 20TH CENTURY PHENOMENON BY DR ELIZABETH DOWSETT

'The body, its systems (such as the gastrointestinal system, the muscular system, the endocrine system, the cardiovascular and vascular systems) and its organs are dependent and their actions largely controlled by the brain. If the brain is physiologically injured, then so is the body. Depending upon which parts of the brain are physiologically injured different parts of the body will also be caused to malfunction.'
DR BYRON HYDE 2006

'It is a fact that the majority of M.E. patients are not in high-stress occupations as the popular press frequently suggests, but are teachers, nurses, physicians, and other health care workers. This group represents those most closely related to infectious illness and those most frequently immunized.'
DR BYRON HYDE

'Following successful immunisation against poliomyelitis in the early 1960s and the removal of 3 strains of polio virus from general circulation in the countries concerned, the related non-polio enteroviruses rapidly filled the vacancy. By 1961, the prevalence of diseases (such as viral meningitis) caused by these agents soared to new heights. In the mid 1980s, the incidence of M.E. had increased by some seven times in Canada and the UK, while in the USA a major outbreak at Lake Tahoe (wrongly ascribed at first to a herpes virus) led to calls for a new name and new definition for the disease, more descriptive of herpes infection. This definition based on "fatigue" (a symptom common to hundreds of diseases and to normal life, but not a distinguishing feature of Myalgic Encephalomyelitis) was designed to facilitate research funded by the manufacturers of new anti-herpes drugs.'
DR ELIZABETH DOWSETT

'Under epidemic and primary M.E. there is no consensus as to the viral or infectious cause. Much of this lack of consensus may be due to the non-separation of acute onset from gradual onset patients in the M.E. and CFS groups of patients. Primary M.E. is always an acute onset illness.

Doctors A. Gilliam, A. Melvin Ramsay and Elizabeth Dowsett, John Richardson, W.H. Lyle, Elizabeth Bell of Ruckhill Hospital, James Mowbray of St Mary's and Peter Behan all believed that the majority of primary M.E. patients fell ill following exposure to an enterovirus (Poliovirus, ECHO, Coxsackie and the numbered viruses are the significant viruses in this group).

I share this belief. Unfortunately, it is very difficult to recover polio and enteroviruses from live patients. Dr James Mowbray developed a test that demonstrated enterovirus infection in many M.E. patients but I do not believe he qualified his patients by acute or gradual onset type of illness. In my tests in Ruckhill Hospital in Glasgow, I found confirmation of enteroviral infection only in acute onset patients and not in any gradual onset [ie. 'CFS'] patients.

Few physicians realize that almost all cases of poliovirus recovered from poliomyelitis victims came from cadavers. At the very least, these enteroviruses must be recovered from patients during their onset illness and this has rarely been done. An exception is in the case of the Lancashire epidemic where Dr W. H. Lyle's investigation recovered ECHO enterovirus.

Recent publications have also identified the fact that enteroviruses are one of the most likely causes of M.E. If this belief is correct, many if not most of the M.E. illnesses could be vanquished by simply adding essential enteroviral genetic material from these enteroviruses to complement polio immunization.'
THE NIGHTINGALE DEFINITION OF M.E. BY DR BYRON HYDE 2006

'In 1948, the year in which polio viruses 1, 2, 3 were first grown in the cells of human or animal tissues Coxsackie viruses (typical of some 12 species of "non polio" enteroviruses with potential to attack neurological tissues in the same way as polio viruses 1-3) were isolated during an epidemic of poliomyelitis in Coxsackie, New York State, USA from 2 boys with typical symptoms. In 1955, an epidemic of neurological illness (later called Myalgic Encephalomyelitis) affected some 300 members of the staff at the Royal Free Hospital, London causing encephalitis accompanied by paresis but without permanent paralysis.

However, many nurses remained chronically disabled and unfit for strenuous hospital duties for the rest of their lives. Some 15 years after this outbreak (1970) two psychiatrists (McEvedy & Beard) studied the case histories of these sufferers and, without seeing or examining a patient, declared the whole incident to be an example of "mass hysteria". Unfortunately, this opinion was to have a profound affect upon the fate of future generations of sufferers from the same illness in the 1980-1989 pandemic. As for McEvedy, he went

on to declare an outbreak of "winter vomiting disease", affecting a boy's school, to be another example of "mass hysteria". By this time further advances in technology (e.g. the electron microscope) clearly indicated a common virus infection to be at fault.'
THE IMPACT OF PERSISTENT ENTEROVIRAL INFECTION BY DR ELIZABETH DOWSETT

'The failure to agree on firm diagnostic criteria has distorted the data base for epidemiological and other research, thus denying recognition of the unique epidemiological pattern of myalgic encephalomyelitis.'
DR A. MELVIN RAMSAY

'We are indebted to Dr Ramsay, an outstanding infectious disease specialist who devoted much effort to the investigation of our disease from the time that he was confronted with the epidemic at the London hospitals in the 1950's. Dr. Ramsay's fame and standing are no accident and we can see that his descriptions of what make this disease unique are accurate and Ramsay's M.E. is the same disease we have today. It is clear that attempts at confusion and name changes would serve to obscure its history and also its origins. So we must never forget Ramsay. The worldwide epidemic we have today is the same disease that Ramsay encountered many years ago.'
THE COMMITTEE FOR JUSTICE AND RECOGNITION FOR M.E.

'Despite the claims of some psychiatrists, it is not true that there is no evidence of inflammation of the brain and spinal cord in M.E.; there is, but these psychiatrists ignore or deny that evidence. It is true that there is no evidence of inflammation of the brain or spinal cord in states of chronic fatigue or 'tiredness.''
THE TERMINOLOGY OF M.E. & CFS BY PROFESSOR MALCOLM HOOPER

ON THE LACK OF FUNDING GIVEN TO LEGITIMATE M.E. RESEARCH, DR BYRON HYDE WRITES: 'Without heed, we are sitting on the edge of a cliff, waiting for disaster. For many sufferers of M.E. that disaster is already here, and few are listening.'

'There is a principle which is a bar against all information, which is proof against all argument, and which cannot fail to keep man in everlasting ignorance. That principle is condemnation without investigation.'
WILLIAM PALEY (1743-1805)

Additional HFME resources available online

Additional HFME papers available on the HFME website include the following:

- A CBT/GET Warning Letter: This letter is for M.E. patients that have tried everything to avoid being forced to participate in CBT or GET programs against their will. It may help to send those who are recommending or administering the program/s a warning letter such as this one. This letter firmly informs those involved *of the scientific facts,* and warns them that *you or your family will sue if you are harmed or killed by these unethical and unscientific interventions.*

- Anaesthesia and M.E.

- Are we just 'marking time?' This paper looks at the problems with the flawed approach contending that 'until we have a unique test for M.E. and more research, we can't expect anything to change'. Unfortunately this approach to M.E. advocacy is popular with some. This paper asks the question: why are we waiting to act when tests for M.E. exist RIGHT NOW, and the need for activism and action is so very urgent?

- A warning on 'CFS' and 'ME/CFS' research and advocacy (co-authored by Lesley Ben): Myalgic Encephalomyelitis and 'Chronic Fatigue Syndrome' are not synonymous terms. The overwhelming majority of research on 'CFS' or 'CFIDS' or 'ME/CFS' or 'CFS/ME' or 'ICD-CFS' does not involve M.E. patients and is not relevant *in any way* to M.E. patients. However, if the M.E. community were to reject all 'CFS' labelled research as *only* relating to 'CFS' patients (including research which describes those abnormalities/characteristics unique to M.E. patients), this may support the myth that 'CFS' is just a 'watered down' definition of M.E. and that M.E. and 'CFS' are virtually the same thing and share many characteristics. A very small number of 'CFS' studies refer in part to people with M.E. but it may not always be clear which parts refer to M.E. This paper includes a checklist to help readers assess the relevance of individual 'CFS' studies to M.E. (if any) and explains some of the problems with this heterogeneous and skewed research.

- M.E. is not fatigue, or 'CFS': This paper explains why M.E. is not defined by mere 'chronic fatigue' and why M.E. and 'CFS' are not synonymous terms, and why a diagnosis of 'CFS' based on any of the definitions of 'CFS' can only ever be a *mis*diagnosis.

- M.E.: The medical facts: A purely medical overview of the illness including detailed research findings.

- Practical tips for living with M.E. plus

    o Tips for coping with M.E. emotionally

    o Tips on resting for M.E. patients

    o Tips for M.E. patients that are parents

    o Assisting the M.E. patient in the use of computers and technology

- Problems with some M.E. (or 'CFS' 'CFIDS' or 'ME/CFS' etc.) advocacy groups: This paper looks at the problems with our advocacy groups, why so many of our groups are not engaged in useful advocacy and what we can do to help change this. This paper is also available as an animated video and in audio format. See the Audio and Video page of the website for details.

- Problems with the so-called "Fair Name" campaign: This paper explains why it is in the best interests of all patient groups involved to reject and strongly oppose this misleading and counter-productive proposal to rename 'CFS' as 'ME/CFS'

- Problems with the use of 'ME/CFS' by M.E. advocates: This paper looks at why it is not in the best interests of M.E. patients and advocates to use, or support the use of, 'ME/CFS.'

- Putting research and articles into context: Because of the politics and financial interests involved in M.E. research it is vitally important that before you read anything about the illness, you understand the context in which it was written. This paper looks at the scientific facts which easily disprove the propaganda surrounding M.E. in so many research studies and articles.

- Smoke and Mirrors: This fully referenced paper looks at the lack of evidence (and the financial and political motivations) behind the fraudulent 'behavioural' model of M.E. and the use and physical effects of interventions such as CBT and GET for patients with M.E. It also outlines a strategy for the resolution of the confusion caused by the 'CFS' disease category. This text forms the introduction to a 100+ page CBT and GET database.

- Testing for M.E.: This paper provides an overview of some of the series of tests which can be done to help confirm a suspected M.E. diagnosis.

- Testing for M.E.: Plan D: Despite the existence of reliable tests for M.E. as described in the 'Testing for M.E.' paper, the reality is that many people who suspect they have M.E. do not have access to the appropriate tests or to doctors who are knowledgeable enough to make a correct diagnosis. This paper discusses the ways in which patients seek a diagnosis, and describes a 'Plan D' for patients who are forced to diagnose (and to some extent, test) themselves.

- The effects of CBT and GET on patients with M.E. No evidence exists to show that CBT or GET are appropriate, safe or helpful for patients with M.E. This paper examines the dangerous physical effects of CBT (psychotherapy) and GET (exercise) on patients with M.E.

- The importance of avoiding overexertion in M.E.

- The misdiagnosis of CFS: None of the definitions of 'CFS' defines M.E., so what do they define? This paper examines what a diagnosis of 'CFS' actually means. See also: A misdiagnosis letter for 'CFS' misdiagnosed (non-M.E.) patients and Where to after a 'CFS' (mis)diagnosis? Every diagnosis of 'CFS' is a misdiagnosis, so where does that leave you and what should you do if a 'CFS' diagnosis is the only diagnosis you have? This paper is designed not for people with M.E., but for everyone who *doesn't* have M.E. but has been given a misdiagnosis of 'CFS' based on any of the 'CFS' definitions and so been denied appropriate diagnosis and treatment for their illness.

- The myths about M.E.

- The comprehensive M.E. symptom list

- The WHO ICD in relation to M.E. and 'CFS'

- Treating M.E.: The basics

  Sections include:

  - Treating and living with M.E.: Overview

  - Important notes on using the HFME's treatment information

  - Symptom-based management vs. deep healing in M.E.

  - Recognising and managing healing reactions in M.E.

  - What if vitamin/mineral/protocol 'x' didn't work for me?

  - Why research and try treatments when some groups claim an M.E. cure is coming soon?

  - Finding a good doctor when you have M.E.

  - A quick start guide to treating and improving M.E. with aggressive rest therapy, diet, toxic chemical avoidance, medications, supplements and vitamins

  - Deep healing in M.E.: An order of attack

  - Food as medicine and M.E.

  - Treating M.E. in the early stages

- What it feels like to have M.E.: A personal M.E. symptom list and description of M.E.

- Who benefits from 'CFS' and 'ME/CFS'? For whose benefit was 'Chronic Fatigue Syndrome' created, and why is it so heavily promoted despite its utter lack of scientific credibility? Who benefits from the artificial 'CFS' construct? Who benefits from Myalgic Encephalomyelitis and 'CFS' being mixed together through unscientific concepts such as 'CFS/ME' and 'ME/CFS' and Myalgic 'Encephalopathy'? Who benefits from ensuring that the facts of M.E. remain ignored, obscured and hidden in plain sight? The short and simple answer to this question is medical insurance companies. This paper examines the involvement of medical insurance companies in the creation of 'CFS' and in the promotion and dominance of the 'CFS' concept today.

- Why care about M.E.? If you aren't personally affected as yet, why should you care about Myalgic Encephalomyelitis and want to help patients achieve the same basic rights as those with similar neurological diseases such as Multiple Sclerosis? This paper explains why you should care about whether or not the fictional disease category of 'CFS' is abandoned.

- Why the disease category of 'CFS' must be abandoned M.E. and 'CFS' are not the same. This paper discusses why renaming, refining or sub-grouping 'CFS' cannot work and why 'CFS' must be abandoned.

- XMRV, 'CFS' and M.E. by Sarah Shenk: There has been much publicity lately over a small research study that claims to have made a connection between "CFS," or "ME/CFS", and a retrovirus. While many are touting this as a huge medical breakthrough, or even claiming that 'our battle is over,' there is a group of doctors, scientists, patients and advocates who question why this research is being said to apply to M.E., when the patients studied merely qualified for a 'CFS' misdiagnosis. They view this information and the claims made about it with tempered enthusiasm or even significant concern. See also: International M.E. expert disputes that 'CFS' XMRV retrovirus claim has relevance to M.E. patients

Additional HFME website pages:

- The HFME M.E. memorial list
- M.E. advocacy and 'CFS' advocacy are not the same
- Support groups
- M.E. book reviews
- M.E. quotes
- Translations
- Audio and video
- Research & articles

  Sections include:

  - General articles and research overviews
  - Cardiac and cardiovascular research
  - Exercise research
  - Mitochondrial muscle research and general muscle research
  - Metabolic research
  - Neurological and cognitive research
  - Immune system research
  - General viral research, enteroviral and post-Polio research
  - Articles exposing the truth behind scams (often aimed at M.E. patients) such as the Lightning Process, Reverse Therapy, Mickel Therapy and EFT .
  - The outbreaks (and infectious nature) of M.E.
  - On the name Myalgic Encephalomyelitis
  - M.E. fatalities
  - Articles sorted by author (including Drs Hyde and Dowsett, the late Drs Ramsay and Richardson, and many more)

The grey ribbon for M.E.

WHO ICD G.93.3

For information on M.E. visit www.hfme.org

Additional HFME websites:

- The HFME YouTube channel A collection of all the original HFME advocacy videos.

- The HFME Facebook group

- The HFME Cafepress store A store featuring shirts, mugs and stickers with the HFME logo. Part proceeds from Cafepress sales benefits the work of the HFME.

- The HFME Amazon.com booklist A list of recommended books for patients, friends, carers and doctors.

- The HFME Yahoo.com support and chat group for M.E. patients This is a group (run by Jodi Bassett) where people with Myalgic Encephalomyelitis can meet other people with the same illness for advice, support and new friendships. M.E. carers are also welcome.

  This is also a group that cares about challenging the myths about 'CFS' and ME and isn't prepared to tolerate the abuse and neglect of M.E. patients. We know that knowledge is power and that debunking these myths is important. The group also contains a penpals section. (Conditions apply to membership of this group.)

- The HFME Yahoo.com support and chat group for severely affected M.E. patients

- The HFME Yahoo.com discussion group This unlisted group is for those who want to play a part in the setting up of HFME, and participate in various HFME projects and discussions of projects. Please contact Jodi Bassett for more information. (Conditions apply to membership of this group.)

- The 'A Hummingbirds' Guide' website is now a personal art website for Jodi Bassett.

How to access this information

To view or download any of these additional HFME papers, please visit the www.hfme.org website and click on either the 'Information Guides' or 'Document Downloads' links, or any of the links on the navigation bar. All papers can be downloaded for free from the website in Word or PDF format, and some papers can also be downloaded in a printable leaflet or business card format for easy redistribution.

To view any of the HFME's web pages or get links for the additional websites listed, please visit the 'Site map' link on the navbar on the www.hfme.org website. New papers are added to the HFME website at least every few months and several new HFME books are being planned for release in 2012 and beyond. To read about new additions to the site, please visit the 'What's new' page on the website or sign up for the free monthly HFME e-newsletter via the website.

'A new idea is first condemned as ridiculous and then dismissed as trivial, until finally, it becomes what everybody knows.'
WILLIAM JAMES (1842-1910)

AFTERWORD

Having Myalgic Encephalomyelitis (M.E). is a traumatic and life-shattering experience and so M.E. patients need all the help they can get. Thank you for taking the time to learn some of the facts of M.E.

There are a lot of shocking and disturbing facts in this book and learning all of them for the first time can be quite overwhelming or emotionally draining. But there is a bright side! By putting a little bit of what you have learned into action you can really improve the life of the person you know or love that has M.E., which is just wonderful, and no small thing.

I'd like to help further, what should I do?

If you feel you are up to trying to help people with M.E. on a larger scale, please join us in helping to spread the facts about M.E. Unlike people with HIV/AIDS, people with M.E. do not have an initial period of their illness where they are only mildly affected. M.E. is severely disabling even in the first week of illness. People with M.E. are almost all far too ill to stage protests, rallies or marches.

Many with M.E. cannot even read enough to be able to understand what is happening, and are not aware that high quality scientific information on M.E. exists, or that supporting the various 'CFS' and 'ME/CFS' faux 'advocacy' groups is counter-productive in the extreme.

What we need is people power! Educated people power!

People from all over the world need to stand up for the truth about M.E.; individual physicians, journalists, politicians, human rights campaigners, patients, families and friends of patients, carers and the general public. That is the only way change will occur; through education and insistence on reform.

People with M.E. have only a tiny minority of the medical, scientific, legal and other potentially supporting professions on their side. M.E. is an infectious neurological disease that is not going away. We must stand together and demand action from our governments, media and medical bodies.

The insurance companies profiting from 'CFS' are acting without regard for the suffering and avoidable deaths they are causing. These groups are acting criminally. There are powerful and immensely wealthy vested interest groups out there which will continue to fight the truth every step of the way, but we have science, reality and ethics on our side and that is also very powerful. However, for this to be of any use to us, we must first make ourselves aware of the facts and then *use them.* Knowledge is power.

Please help us to spread the truth about Myalgic Encephalomyelitis and expose the lie of 'CFS.'

You can also help by NOT supporting the bogus concepts of 'CFS,' 'ME/CFS,' 'subgroups of ME/CFS,' 'CFS/ME,' 'CFIDS' and Myalgic 'Encephalopathy.'

Do not give public or financial help or support to groups which promote these harmful and unscientific concepts or which equate M.E. with 'CFS.'

Some practical ways to help include sending links to HFME papers to friends and family, regularly posting HFME links on websites containing inaccurate information about the disease, distributing the free printable HFME leaflets to your friends and family or to all the doctors in your area, buying a few copies of HFME books or books by Dr Byron Hyde for your local library (or any other groups or individual that may benefit by reading them), or positively reviewing your favourites of these books online. You might also sign the HFME guestbook online, or offer to help HFME contributors with services such as translation or editing.

Please see the HFME website for further information on our M.E. advocacy campaigns and all the big and small ways you can help HFME and M.E. patients generally. We're also open to any new ideas you might have. Every contribution helps.

The abuse and neglect of so many severely ill people on such an industrial scale is truly inhuman and has already gone on for far too long.

People with M.E. desperately need your help.

As anthropologist Margaret Mead famously said, '*Never doubt that a small group of thoughtful, committed citizens can change the world. Indeed, it's the only thing that ever has.*'